PROPERTIES AND USES OF MICROEMULSIONS

Edited by **Desiree Nedra Karunaratne,**
Geethi Pamunuwa
and **Udayana Ranatunga**

Properties and Uses of Microemulsions
http://dx.doi.org/10.5772/63187
Edited by Desiree Nedra Karunaratne, Geethi Pamunuwa and Udayana Ranatunga

Contributors

H Scott Fogler, Christopher Fredd, Mark Hoefner, Valery Tarasov, Makcolm Forbes, Tajamal Hussain, Rabia Batool, Magnus Bergström, Taisuke Banno, Kouichi Asakura, J. C. Mejuto, Lacramioara Ochiuz, Desiree Nedra Karunaratne

Published by InTech

Janeza Trdine 9, 51000 Rijeka, Croatia

Publishing Process Manager Edi Lipovic
Technical Editor SPi Global
Cover InTech Design team

First published May, 2017
Printed in Croatia
Legal deposit, Croatia: National and University Library in Zagreb

Additional hard copies can be obtained from orders@intechopen.com

Properties and Uses of Microemulsions , Edited by Desiree Nedra Karunaratne, Geethi Pamunuwa and Udayana Ranatunga
p. cm.
Print ISBN 978-953-51-3171-7
Online ISBN 978-953-51-3172-4

For EU product safety concerns:
IN TECH d.o.o., Prolaz Marije Krucifikse Kozulić 3, 51000 Rijeka, Croatia,
info@intechopen.com or visit our website at intechopen.com.

Meet the editors

Desiree Nedra Karunaratne, Professor of Chemistry, University of Peradeniya, obtained a BSc degree (Chem. Hons.) from the University of Colombo, Sri Lanka, and a PhD degree from the University of British Columbia, Canada. She widely researches in areas such as nanotechnology applications of polymer nanoparticles and liposomes for delivery of active ingredients in pharmaceutics, cosmetics, and nutraceuticals in food; effect of liquid crystals in emulsion-based skin formulations; and structural determinations of bacterial polysaccharides and searches for bioactive natural products. She has over 50 publications in peer-reviewed journals, 7 patents, 11 book chapters, and 96 presentations at conferences. Her editorial experience spreads over scientific books as well as the scientific magazine of the University of Peradeniya.

Geethi Pamunuwa obtained her BSc degree (Chem. Hons.) from the University of Peradeniya in 2006, MA degree (Chemistry) from Wayne State University, USA, in 2010, and PhD degree (Chemical Sciences) from the University of Peradeniya, Sri Lanka, in 2015. Her research interests include transdermal delivery of bioactive agents, liposomes, polymer nanoparticles, food formulation and preservation, and biopesticides. She currently works as a senior lecturer in food science and technology at the Department of Horticulture and Landscape Gardening, Wayamba University of Sri Lanka. She has published 5 articles in peer-reviewed journals and made 7 presentations at various conferences.

Udayana Ranatunga obtained his bachelor's degree from the University of Peradeniya in 2006 and his doctoral degree in Chemistry from the University of Texas at Dallas in 2012. His research is focused on the theoretical description of interfaces and nanomaterials and is trained in computational chemistry techniques. He has several publications to his credit and is currently working as a senior lecturer of chemistry, at the Department of Chemistry, University of Peradeniya, Sri Lanka.

Contents

Preface

In recent years, colloidal systems have taken a turn toward micro- and nanoscale from the conventional macroemulsions. Microemulsions cover a large area of investigative research. The smaller size of the droplets and the transparency of the emulsion contribute to increased interest in applications of these emulsions. The introductory chapter takes a broad overview of the many uses of microemulsions and the vast areas in which they are applicable. From applications ranging from coating and textile finishes to lubricants and drug delivery vehicles, their versatility in industrial applications is unmatched.

In the second section of the book, some practical applications limited to the use of microemulsions for delivery of bioactive components, for industrial carbonate reservoir stimulation, and for preparation of nanocatalysts are presented. Recently, microemulsions have been shown to display better delivery properties compared to macroemulsions in the area of drug delivery. Several bioactive compounds have therefore been investigated for the purpose of faster drug delivery and rapid penetration of topical formulations. The main advantage of an emulsion is its ability to create a path to transport drugs through membrane barriers and incorporate drugs into inert lipid vehicles. These microemulsions have inherent characteristics of improving drug solubility, long shelf life, and cost-effectiveness. Their ability to accommodate both hydrophilic and hydrophobic drugs has resulted in improved solubility and thermodynamic stability, which is a dominant advantage. Thus, microemulsion dermal products may be a common medium for administering various herbal extracts and plant oils. To this, Ochiuz and Hortolomei have studied the use of avocado oil as the oil phase, making a change from the traditional use of oils such as olive oil for skin formulations. They have taken advantage of the beneficial properties of avocado oil on the human skin as the basis for its use in their formulation. Erythromycin has been used as the drug to be delivered by the avocado oil system and has successfully characterized and optimized the system for efficient delivery.

A novel application for stimulating carbonate reservoirs is presented by Fredd et al. Generally, carbonate reservoirs trapping oil hydrocarbons are treated by hydrochloric acid, high-pressure methods, or injection of saline water to fracture and form fissures to release the trapped oil. The use of microemulsions prepared with HCl as the dispersed phase is shown to reduce the effective diffusivity, which can increase the depth of stimulation unlike the rapid reaction of HCl with the carbonate, which reduces the depth of stimulation. Therefore, better recovery is possible with the help of microemulsions.

In the third application, Hussain and Batool have employed microemulsions to prepare nanostructured catalysts. MACROBUTTON NoMacro Nanocatalysts with specific shapes, morphology, surface area, size, geometry, and homogeneity can be fabricated with the formation of microemulsions. The size of the nanoparticles can be controlled by several physical parameters leading to the formation of highly monodispersed metal nanoaggregates. Metals such as Au, Pd, Rh, etc. and materials such as silica, alumina, and metal oxides may be prepared by microemulsion techniques to function as nanocatalysts in a wide range of applications.

The morphology of a microemulsion depends on the droplet size and nature of the phases used for emulsion formation. The behavior of the droplets in the system is subject to thermodynamics of self-assembly, the interfacial tension, self-propelled motion, and other physical attractions. Therefore, Section 3 of this volume is devoted to the theoretical and physical parameters predicting microemulsion properties. Tarasov and Forbes have used spin-polarized paramagnetic probes to investigate the molecular mobility of microemulsions. Time-resolved electron paramagnetic resonance (TREPR) spectroscopy has enabled them to study liquid-phase inhomogeneity of spin-correlated radical pairs (SCRPs) and thereby quantitate molecular mobility. L. M. Bergstrom looks at the bending energetics of microemulsions using three bending elasticity constants, namely, spontaneous curvature (H_0), bending rigidity (k_c), and saddle-splay constant (QUOTE). His model relates microemulsion droplet size with polydispersity, interfacial tension, and solubilization capacity.

Since emulsions are dynamic systems, the motion of the oil droplets in the surfactant solution requires understanding. Banno, Toyota, and Asakura studied the transfer of surfactant through the water-oil interface, which results in self-propelled motion of the droplets. By observing the movements using various microscopes, they have tried to unravel the mechanism so that more stable emulsion systems could be developed. The final chapter by Mejuto et al. describes the use of artificial neural networks for predicting how AOT-based microemulsions are influenced by amphiphiles. The predictions have been obtained using models of varying composition. These studies have shown that the physicochemical variables may be identified through prediction models.

Microemulsions may be considered as a novel medium for a variety of applications. Therefore, understanding the theory behind the mechanism is very important for the purpose of improved and newer applications.

Desiree Nedra Karunaratne
Professor in Chemistry,
University of Peradeniya, Sri Lanka

Geethi Pamunuwa
Department of Horticulture and Landscape Gardening,
Faculty of Agriculture and Plantation Management,
Wayamba University of Sri Lanka, Sri Lanka

Udayana Ranatunga
Department of Chemistry,
University of Peradeniya, Sri Lanka

Section 1

Introduction to Microemulsions

Chapter 1

Introductory Chapter: Microemulsions

D. Nedra Karunaratne, Geethi Pamunuwa and
Udayana Ranatunga

Additional information is available at the end of the chapter

http://dx.doi.org/10.5772/intechopen.68823

1. Introduction

An emulsion is a dispersion of droplets of one liquid in a second immiscible liquid. The droplets are termed the dispersed phase, while the second liquid is the continuous phase. To stabilize an emulsion, a surfactant or co-surfactant is added such that the droplets remain dispersed and do not separate out as two phases. Depending on the phase, there are two types of microemulsions: water-in-oil (w/o) and oil-in-water (o/w). As the name implies, water is the dispersed phase in w/o emulsions, whereas oil is the dispersed phase in o/w emulsions [1, 2]. One of the main differences between macroemulsions and microemulsions is that the size of the droplets of the dispersed phase of microemulsions is between 5 and 100 nm, while that of macroemulsions is >100 nm. Microemulsions are thermodynamically stable systems, whereas macroemulsions are kinetically stable systems [2]. Also, microemulsions are translucent and of low viscosity, while macroemulsions are opaque and of relatively high viscosity. Due to these unique properties of microemulsions, these systems have become indispensible in numerous important fields.

2. Characterization

Microemulsions, at the molecular scale, are a finely balanced system where the energetics of entropy and surface energies are opposing each other. The entropy of the system is increased by having a higher number of droplets dispersed, while the surface areas, and correspondingly surface energies are increased with more droplets. Slight changes in the chemical composition or conditions can shift this balance and may, therefore, lead to dramatic changes in the behavior of the system.

The expansion and integration of microemulsions into various applications (see Section 3) have imbued microemulsions with industrial and commercial importance. To optimize formulations and predictive models, adequate characterization is essential. Because microemulsions are essentially a subset of the broader class of emulsions, techniques which pertain to general emulsions also play a part in this regime. However, techniques which are specific to microemulsions have also emerged.

2.1. Phase behavior

Characterization of microemulsions necessarily begins with the elucidation of the phase behavior, where, in the most simple form, oil, water, and surfactant are mixed in different molar ratios, and the formed phases are tracked [3]. A generic phase diagram resulting from such analysis is shown in **Figure 1**. In several regions of the phase diagram, single phases exist in one end where spherical oil droplets are dispersed in water (o/w emulsions) and the other where spherical water droplets are dispersed in oil (w/o emulsions). Between these two extremes, depending on the ratio of components, the droplet shapes may change from spherical to cylindrical, to worm-like micelle, to bicontinuous [4]. Traditionally methods of microscopy, rheology, conductivity, and nephelometry have been used in the study of microemulsions, while instrumental techniques such as dynamic light scattering, neutron scattering, X-ray scattering, electron micrography, nuclear magnetic resonance, electron paramagnetic resonance, and their derivatives are widely used at present. Although detailed treatise of these myriad techniques is outside of the scope of this chapter; here, we briefly discuss the physical properties important to characterize microemulsions, and the techniques able to quantify them.

The phase behavior of the system can be characterized by several techniques. Other than using visual inspection, optical microscopy using polarized light is commonly used [5] to detect the presence of a singular phase, or the presence of separate phases. Simple electrical conductivity (EC) measurements can be used to identify whether the oil/water or both phases are continuous [5, 6]. The EC of a w/o emulsion increases with the addition of water. After the maximum amount of water that water in oil emulsion can hold is exceeded, the emulsion col-

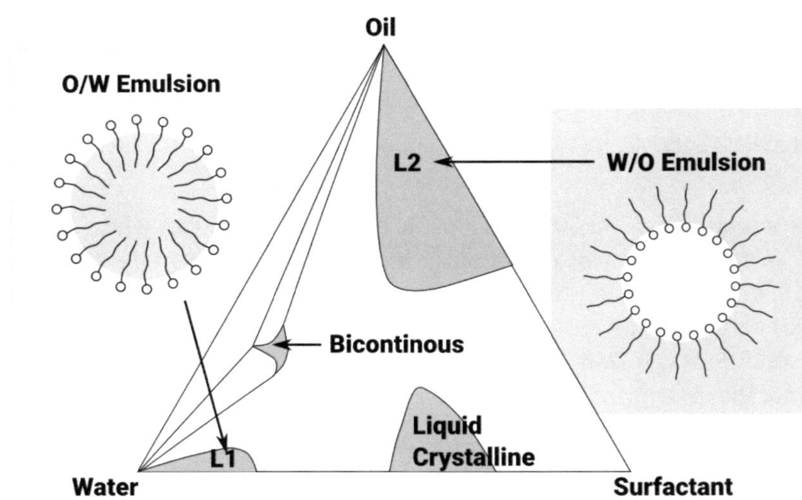

Figure 1. Example of a ternary phase diagram of oil/water/surfactant system. Here, the formation of oil-in-water (o/w) emulsions or water-in-oil (w/o) emulsions is determined by the molar ratio of the components. Adapted from Archarya et al. [3].

lapses and the EC decreases. On the other hand, for o/w emulsions, the EC is steady until the concentration of oil droplets increases and an effectively infinite conductive pathway arises leading to dramatic increases in measurements [6]. Microscopic techniques such as scanning electron microscopy (SEM) or transmission electron microscopy (TEM) have adequate resolution to image individual droplets or the morphology of a larger region. However, dehydration required for conventional sample preparation can severely affect the native structure [3, 7, 8].

2.2. Droplet size

Apart from the existence of the unitary phase, the size of droplets of the dispersed phase of a microemulsion is a primary characteristic, since it determines much of the physical behavior and functionality. Conventionally, the term, 'microemulsion' refers to a thermodynamically stable, isotropic emulsion with a droplet size of 1–100 nm. For macroemulsions, both sound (acoustics, electroacoustics) and visible light (turbidimetry, nephelometry, laser diffraction) can be used to characterize the droplet size. However, for emulsions of which the droplet size is less than 1 μ, the intensity of acoustic attenuation and diffuse light scattering decreases with droplet size. This leads to a useful property that microemulsions are not only transparent but also requires different techniques to characterize them.

In the domain of microemulsions, size (and shape) dependant scattering of visible light, X-Rays, electrons, and neutrons can be interrogated to yield useful information. Dynamic light scattering (DLS or PCS) is a widely used for characterizing particle size and shape anisotropy [3]. Here, the measurement is the fluctuation of scattered laser light intensity, which is related to the Brownian motion of particles (droplets) in the medium. Fitting the autocorrelation of the scattered light intensity to models of particles in specific media can lead to determining the diffusion constant, particle size distribution, and in some cases, shape anisotropy [9–11].

In the case of X-rays, because oil and water scatter X-rays to different extents, small angle X-ray scattering (SAXS) can also be used to obtain the size of the dispersed phase [3, 12]. Neutrons can also be used as the primary beam, and the variation in scattering intensity by different nuclei gives rise to small-angle neutron scattering (SANS). SANS data give access to both the droplet size and dynamic properties; however, the relative cost and data collection time limit the applicability of this technique [3].

Apart from scattering techniques, nuclear magnetic resonance (NMR) and related techniques are widely used in the characterization of microemulsions [3, 13]. Here, both relaxation [14] and self-diffusion experiments [15] are used to probe the emulsion systems [13, 16].

2.3. Droplet size distribution

It follows that most techniques which give a measure of the droplet size of microemulsions will also have access to the size distribution or polydispersity. Scattering techniques such as DLS, time-averaged light scattering, and SANS can characterize size distribution. However, due to the confluence of particle size variation and shape anisotropy, analyzing, and deconvoluting, these data may not always be straightforward. SANS is particularly versatile and can characterize the structure, and interactions of colloidal systems with a wide range of interactions as well as differences in orientation and droplets with non-spherical shapes [17].

2.4. Diffusion constant

Similarly, because most techniques use the motion of aggregates (droplets) to elucidate the sizes of dispersed droplets, light scattering techniques have access to the diffusion rates as well. However, it is important to note that fluorescence correlation spectroscopy (FCS) [3] can be used to measure the diffusion constants (in addition to particle size and size distribution [18]), particularly useful in dilute solutions and other conditions where light scattering techniques fail [3]. However, when the droplet sizes of the emulsions become comparable to the incident beam focus size, the FCS is not reliable leading to the technique being used as a complementary method to DLS [3].

Nuclear magnetic resonance and derivatives of the technique can be used to measure the self-diffusion constants of oil, water and the surfactants in microemulsions [3]. Microemulsion structure is best investigated by NMR relaxation and self-diffusion studies. Bicontinuity in equilibrium microemulsions can be obtained by self-diffusion studies. NMR relaxation is very sensitive to the droplet size but is insensitive to interactions allowing accurate droplet size measurements [13].

2.5. Morphology

Although all of the techniques discussed so far give information about the morphology, electron microscopy provides a robust method to directly visualize the nanoscale structure and morphology of microemulsions. For conventional TEM images, a thin ($\sim 10^2$ nm) specimen is required, while for SEM, deposition and fixation on a solid substrate are usually required. In each of these cases, the sample preparation requires dehydration which, in the case of microemulsions, can severely affect their native structure [3]. To alleviate this problem, cryo-electron microscopy (cryo-EM) techniques have been utilized where emulsions are frozen in their native hydrated states. From this point, preparing samples through freeze-fracture or freeze-etching [7, 8] can reveal rich and unique insight into the systems.

2.6. Rheology

The rheological properties of microemulsions are often crucial in their application because they will affect the processability, and kinetics, and stability under various conditions [19]. Microemulsions show varying rheology depending on the phase point. For example, o/w and w/o microemulsions show Newtonian behavior over a wide range of shears, while the bicontinuous phase may undergo breakage upon medium shear forces, leading to thinning [3]. Although the effect of the molecular structure of emulsions has a large impact on the behavior of microemulsions [20], the characterization techniques are generally the same as their macroscopic counterparts [21, 22].

3. Applications

The applications of microemulsions are plenteous and span in areas including drug delivery, cosmetics, food, fuel, lubricants and coatings, detergents, agrochemicals, analytical chemistry, nanoparticle synthesis, biotechnology, and chemical reactors [23]. Exhibiting a pseudo biphasic behavior, these systems allow solubilization of highly hydrophilic substances in oil-based systems and highly hydrophobic substances in water-based systems. Further, ultralow interfacial tension, the presence of nanosized droplets of dispersed phase, slow release and protection of encapsulated material, and the ability to penetrate through biological membranes are some attributes that make the microemulsions find significant applications in various sectors. A brief description of the applications of microemulsions in selected fields is given below.

3.1. Pharmaceutics

Emulsions are opaque gels or creams in which a drug is dispersed for topical application. The effect of the drug released from the gel depends on the permeability of the drug through the skin barrier. Microemulsions considering the small size of the droplets can serve as better delivery vehicles thereby improving the drug solubility, penetrability, and shelf life [24]. An advantage of these systems is their ability to deliver both hydrophobic and hydrophilic drugs efficiently via o/w or w/o emulsions, respectively [25]. It has been shown that penetrability of hydrophobic drugs is improved by encasing the drug in a lipid vesicle [26]. Therefore, it is apparent that the smaller the size of the droplet, the better the delivery of a hydrophobic drug. Drug diffusion was shown to follow kinetics related to models such as Higuchi model resulting in the slow release of the drug [27]. In this case, the permeation was shown to increase when glycolipids were incorporated into the microemulsion indicating that they could outperform macroemulsions in topical drug delivery. Poorly soluble drugs such as cyclosporine and paclitaxel have shown improved oral bioavailability in microemulsion systems and have been patented along with other drugs such as ritonavir and saquinavir [28].

3.2. Cosmetics

Cosmetics and cosmeceutics currently utilize microemulsion systems and demonstrate the enormous potential of using these systems for various products. Skin care products, hair care products, and perfumes are the main types of microemulsion products available in the market. The surfactants, co-surfactants, and oils used in cosmetic microemulsions are either natural or synthetic. The surfactants are either ionic or nonionic [23]. Bioactive agents, including antioxidants and skin whitening agents, have been incorporated in and delivered to the skin via microemulsion cosmetic products [29, 30]. Interestingly, antioxidant and moisturizing effects of olive oil, which can be utilized as the main ingredient in microemulsions, increase upon incorporation in microemulsions thus making such systems apt for cosmetic applications [31]. In addition to delivering nutrients and increasing moisturizing effects, microemulsions have been identified as promising systems for removing oily make-up cosmetics from the skin [32].

3.3. Food

Numerous attributes of microemulsions render these systems excellent to be used in the food sector. Among such attributes, their ability to protect, slowly release, and enhance the activity of the encapsulated material, and the possibility of formulating microemulsions using edible substances, stand out. According to a recent study, garlic essential oils encapsulated in water-based microemulsions have exhibited antimicrobial activity indicating its potential use in the food industry [33]. Further, the bioactive compounds—crocin, safranal, and picrocrocin—of saffron encapsulated in multiple emulsions have shown enhanced slow release properties and greater stability in gastric conditions [34]. Moreover, microemulsions encapsulating steppogenin have shown to be effective in reducing enzymatic browning of apple juice. Co-encapsulation of vitamin C with steppogenin greatly enhances this antibrowning effect [35]. The number of studies on applications of microemulsions in food is plenteous and is still growing.

3.4. Enhanced oil recovery

Microemulsions are used in enhanced oil recovery, and approximately 20% enhanced oil recovery has been reported. The high interfacial tension between the crude oil and reservoir brine keeps the residual oil in the reservoir. The interfacial tension can be lowered via the preparation or introduction of microemulsions, and thus, this area is actively investigated. A surfactant formulation is injected into the reservoir in the surfactant-polymer flooding process. The surfactants stimulate the formation of a microemulsion in the porous reservoir between reservoir brine and crude oil, which reduces the interfacial tension between the two. Hence, the oil recovery is enhanced [36]. A more recent trend is to utilize more cost-effective microbes to produce microbial products including biosurfactants in place of chemical mixtures [37, 38]. Also, numerous studies have been conducted to evaluate the use of ionic liquids as green chemicals in place of surfactants in microemulsions in enhancing oil recovery [39].

3.5. Fuels

Microemulsions have been used as fuels with many attractive properties. These fuels are used to decrease the emission rates of gases such as nitrogen oxides and carbon monoxide, and particles (soot) [40]. Although alcohols frequently used in microemulsion biofuels decrease the cetane number. The incorporation of cetane improvers has significantly increased the cetane number thus improving the properties of microemulsion fuels [41]. Moreover, water in the microemulsion-based fuel reduces the combustion temperature and heat released. Due to the increased surface area, the air-fuel contact is improved. Overall, the fuel efficiency is improved as microemulsion-based fuels are used [42].

3.6. Lubricants, cutting oils and corrosion inhibitors

Microemulsion systems are frequently used as lubricants. Microemulsions prepared using ionic liquids and copper nanoparticles are some recent advances in this field [43, 44]. As cutting oils, microemulsions serve as lubricants and absorbers of the heat of friction [45]. As corrosion inhibitors, microemulsions may show many mechanisms of action. The corrosion causing factors may be soluble in the microemulsion so that those factors may be unavailable for the metal surface. Also, the hydrophobic coating on the metal may prevent corrosion [46].

3.7. Coatings and textile finishing

Microemulsion-based resins exhibit superior properties than solvent-based resins. Mainly, flammability hazards, health problems, and pollution problems are much less associated with microemulsion-based coatings. Further, microemulsion coatings are better than emulsion coatings with respect to scrub resistance, stain resistance and color intensity [23, 47, 48]. The suitability of microemulsions in textile finishing has also been demonstrated by many researchers. In fact, microemulsions have shown better properties than both conventional textiles finishing aids and normal emulsions [23, 49].

3.8. Nanoparticle synthesis

Microemulsions have been commonly used as means of preparing nanoparticles. Recently, bimetallic nanocatalysts comprising Cu and Ni were formed using reserved microemulsions where they altered the microemulsion composition to obtain different sizes of bimetallic particles [50]. Narrow size distribution and regular shape are two other important attributes of metallic nanoparticles synthesized using the reversed microemulsion method [51–53]. In addition to inorganic nanoparticles, organic nanoparticles such as cholesterol and rhovanil have been prepared successfully using microemulsions. Also, the properties of nanoparticles may, in certain instances, be modulated by changing the physicochemical properties of microemulsions [54].

Author details

D. Nedra Karunaratne[1]*, Geethi Pamunuwa[2] and Udayana Ranatunga[1]

*Address all correspondence to: nedrak@pdn.ac.lk

1 Department of Chemistry, Faculty of Science, University of Peradeniya, Peradeniya, Sri Lanka

2 Department of Horticulture and Landscape Gardening, Faculty of Agriculture and Plantation Management, Wayamba University of Sri Lanka, Makandura, Gonawila, Sri Lanka

References

[1] Hoar TP, Schulman JH. Transparent water in oil dispersions: The oleopathic hydromicelle. Nature. 1943;**152**:102–103. DOI: 10.1038/152102a0

[2] Moulik SP, Rakshit AK. Physicochemistry and applications of microemulsions. Journal of Surface Science and Technology. 2006;**22**(3–4):159–186. DOI: 10.18311/jsst/2006/1965

[3] Acharya DP, Hartley PG. Progress in microemulsion characterization. Current Opinion in Colloid and Interface Science. 2012;**17**:274–280. DOI: 10.1016/j.cocis.2012.07.002

[4] Najjar R. Microemulsions: An Introduction to Properties and Applications. Intech; 2012; pp. 1–30. DOI: 10.5772/2300

[5] Lutz R, Aserin A, Wachtel EJ, Ben-Shoshan E, Danino D, Garti N. A study of the emulsified microemulsion by SAXS, Cryo—TEM, SD—NMR, and electrical conductivity, A

study of the emulsified microemulsion by SAXS, Cryo-TEM, SD-NMR, and electrical conductivity. Journal of Dispersion Science and Technology. 2007;**28**:1149–1157. DOI: 10.1080/01932690701525452

[6] Lagues M, Sauterey C. Percolation transition in water in oil microemulsions. Electrical conductivity measurements. The Journal of Physical Chemistry A. 1980;**84**:3503–3508. DOI: 10.1021/j100463a003

[7] Strey R. Microemulsion microstructure and interfacial curvature. Colloid and Polymer Science. 1994;**272**:1005–1019. DOI: 10.1007/BF00658900

[8] Gulik-Krzywicki T, Larsson K. An electron microscopy study of the L2-phase (microemulsion) in a ternary system: Triglyceride/monoglyceride/water. Chemistry and Physics of Lipids. 1984;**35**(2):127–132. DOI: 10.1016/0009-3084(84)90018–5

[9] Muller BW, Muller RH. Particle size distributions and particle size alterations in microemulsions. Journal of Pharmaceutical Sciences. 1984;**73**:919–922. DOI: 10.1002/jps.2600730714

[10] Papadimitriou V, Pispas S, Syriou S, Pournara A, Zoumpanioti M, Sotiroudis TG, Xenakis A. Biocompatible microemulsions based on limonene: Formulation, structure, and applications. Langmuir. 2008;**24**(7):3380–3386. DOI: 10.1021/la703682c

[11] Aubery C, Solans C, Prevost S, Gradzielski M, Sanchez-Dominguez M. Microemulsions as reaction media for the synthesis of mixed oxide nanoparticles: Relationships between microemulsion structure, reactivity, and nanoparticle characteristics. Langmuir. 2013;**29**:1779–1789. DOI: 10.1021/la303817w

[12] Shimobouji T, Matsuoka H, Ise N, Oikawa H. Small-angle x-ray scattering studies on nonionic microemulsions. Physical Review A. 1989;**39**:4125–4131. DOI: 10.1103/PhysRevA.39.4125

[13] Lindman B, Olsson U. Structure of microemulsions studied by NMR. Berichte Der Bunsen Gesellschaft Fuer Physikalische Chemie. 1996;**100**:344–363. DOI: 10.1002/bbpc.19961000321

[14] Kataoka H, Ueda T, Ichimei D, Miyakubo K, Eguchi T, Takeichi N, Kageyama H. Evaluation of nanometer-scale droplets in a ternary o/w microemulsion using SAXS and 129Xe NMR. Chemical Physics Letters. 2007;**441**(1–3):109–114. DOI: 10.1016/j.cplett.2007.04.088

[15] Law SJ, Britton MM. Sizing of reverse micelles in microemulsions using NMR measurements of diffusion. Langmuir. 2012;**28**:11699–11706. DOI: 10.1021/la300796u

[16] Furó I. NMR spectroscopy of micelles and related systems. Journal of Molecular Liquids. 2005;**117**(1–3):117–137. DOI: 10.1016/j.molliq.2004.08.010

[17] Schubert KV, Strey R. Small angle neutron scattering from microemulsions near the disorder line in water/formamide-octane-C_iE_j systems. The Journal of Chemical Physics. 1991;**95**(12):8532–8545. DOI: 10.1063/1.461282

[18] Pal N, Verma SD, Singh MK, Sen S. Fluorescence Correlation Spectroscopy: An efficient tool for measuring size, size-distribution and polydispersity of microemulsion droplets in solution. Analytical Chemistry. 2011;**83(20)**:7736–7744. DOI: 10.1021/ac2012637

[19] Derkach SR. Rheology of emulsions. Advances in Colloid Interface Science. 2009;**151:**1–23. DOI: 10.1016/j.cis.2009.07.001

[20] Dawson K, Pa G. Connection of microstructure to rheology in a microemulsion model. Physical Review E. Statistical Physics, Plasmas, Fluids, and Related Interdisciplinary Topics. 1996;**54:**1669–1682

[21] Kantaria S, Rees GD, Lawrence MJ. Gelatin-stabilised microemulsion-based organogels: Rheology and application in iontophoretic transdermal drug delivery. Journal of Controlled Release. 1999;**60**(2-3):355–365. DOI: 10.1016/S0168-3659(99)00092-9

[22] Lapasin R, Grassi M, Coceani N. Effects of polymer addition on the rheology of o/w microemulsions. Rheologica Acta. 2001;40(2):185–192. DOI: 10.1007/s003970000151

[23] Paul BK, Moulik SP. Uses and applications of microemulsions. Current Science. 2001; **80**(8):990–1001

[24] Kogan A, Garti N. Microemulsions as transdermal drug delivery vehicles. Advances in Colloid and Interface Science. 2006;**123–126**:369–385. DOI: 10.1016/j.cis.2006.05.014

[25] Lawrence MJ, Rees GD. Microemulsion-based media as novel drug delivery systems. Advanced Drug Delivery Reviews. 2000;**45**(1):89–121. DOI: 10.1016/S0169-409X(00)00103-4

[26] Karunaratne DN, Dassanayake AC, Pamunuwa KMGK, Karunaratne V. Improved skin permeability of *dl*-α-tocopherol in topical macroemulsions. International Journal of Pharmacy and Pharmaceutical Sciences. 2014;**6**(6):53–57

[27] Premarathne EPN, Karunaratne DN, Perera ADLC. Controlled release of diclofenac sodium in glycolipid incorporated micro emulsions. International Journal of Pharmaceutics. 2016;**511**(2):890–898. DOI: 10.1016/j.ijpharm.2016.07.047

[28] Talegaonkar S, Azeem A, Ahmad FJ, Khar RK, Pathan SA, Iqbal Z. Microemulsions: A novel approach to enhanced drug delivery. Recent Patents on Drug Delivery & Formulation. 2008;**2**:238–257

[29] Montenegro L. Nanocarriers for skin delivery of cosmetic antioxidants. Journal of Pharmacy and Pharmacognosy Research. 2014;**2**(4):73–92

[30] Boonme P, Junyaprasert VB, Suksawad N, Songkro S. Microemulsions and nanoemulsions: Novel vehicles for whitening cosmeceuticals. Journal of Biomedical Nanotechnology. 2009;**5**(4):373–383. DOI: 10.1166/jbn.2009.1046

[31] Chaiyana W, Leelapornpisid P, Phongpradist R, Kiattisin K. Enhancement of antioxidant and skin moisturizing effects of olive oil by incorporation into microemulsions. Nanomaterials and Nanotechnology. 2016;**6**:1–8. DOI: 10.1177/1847980416669488

[32] Kim EJ, Kong BJ, Kwon SS, Jang HN, Park SN. Preparation and characterization of W/O microemulsion for removal of oily make-up cosmetics. International Journal of Cosmetic Science. 2014;**36**(6):606–612. DOI: 10.1111/ics.12163

[33] El-Sayed HS, Chizzola R, Ramadan AA, Edris AE. Chemical composition and antimicrobial activity of garlic essential oils evaluated in organic solvent, emulsifying, and self-microemulsifying water based delivery systems. Food Chemistry. 2017;**221**:196–204. DOI: 10.1016/j.foodchem.2016.10.052

[34] Esfanjani AF, Jafari SM, Assadpour E. Preparation of a multiple emulsion based on pectin-whey protein complex for encapsulation of saffron extract nanodroplets. Food Chemistry. 2017;**221**:1962–1969. DOI: 10.1016/j.foodchem.2016.11.149

[35] Tao J, Zhu Q, Qin F, Wang M, Chen J, Zheng Z-P. Preparation of steppogenin and ascorbic acid, vitamin E, butylated hydroxytoluene oil-in-water microemulsions: Characterization, stability, and antibrowning effects for fresh apple juice. Food Chemistry. 2017;**224**:11–18. DOI: 10.1016/j.foodchem.2016.12.045

[36] Bera A, Mandal A. Microemulsions: A novel approach to enhanced oil recovery: A review. Journal of Petroleum Exploration and Production Technology. 2015;**5**(3):255–268. DOI: 10.1007/s13202-014-0139-5

[37] Gassara F, Suri N, Voordouw G. Nitrate-mediated microbially enhanced oil recovery (N-MEOR) from model upflow bioreactors. Journal of Hazardous Materials. 2017;**324**(A):94–99. DOI: 10.1016/j.jhazmat.2015.12.039

[38] Sivasankar P, Kumar GS. Influence of pH on dynamics of microbial enhanced oil recovery processes using biosurfactant producing Pseudomonas putida: Mathematical modelling and numerical simulation. Bioresource Technology. 2017;**224**:498–508. DOI: 10.1016/j.biortech.2016.10.091

[39] Bera A, Belhaj H. Ionic liquids as alternatives of surfactants in enhanced oil recovery—A state-of-the-art review.Journal of Molecular Liquids. 2016;**224**(A):177–188. DOI: 10.1016/j.molliq.2016.09.105

[40] Kazerooni H, Rouhi A, Khodadadi AA, Mortazavi Y. Effects of combustion catalyst dispersed by a novel microemulsion method as fuel additive on diesel engine emissions, performance, and characteristics. Energy Fuels. 2016;**30(4)**:3392–3402. DOI: 10.1021/acs.energyfuels.6b00004

[41] Atmanli A. Effects of a cetane improver on fuel properties and engine characteristics of a diesel engine fueled with the blends of diesel, hazelnut oil and higher carbon alcohol. Fuel. 2016;**172**:209–217

[42] Patel PD, Lakdawala A, Chourasia S, Patel RN. Bio fuels for compression ignition engine: A review on engine performance, emission and life cycle analysis. Renewable and Sustainable Energy Reviews.2016;**65**:24–43. DOI: 10.1016/j.rser.2016.06.010

[43] Wang A, Chen L, Jiang D, Yan Z. Vegetable oil-based ionic liquid microemulsions and their potential as alternative renewable biolubricant basestocks. Industrial Crops and Products. 2013;**51**:425–429. DOI: 10.1016/j.indcrop.2013.09.039

[44] Wang A, Chen L, Xu F, Yan Z. In situ synthesis of copper nanoparticles within ionic liquid-in-vegetable oil microemulsions and their direct use as high efficient nanolubricants. Royal Society of Chemistry Advances. 2014;**4**:45251–45257. DOI: 10.1039/C4RA06785A

[45] Bataller H, Lamaallam S, Lachaise J, Graciaa A, Dicharry C. Cutting fluid emulsions produced by dilution of a cutting fluid concentrate containing a cationic/nonionic surfactant mixture. Journal of Materials, Processing Technology. 2004;**152**(2):215–220. DOI: 10.1016/j.jmatprotec.2004.03.027

[46] Moura EF, Neto AOW, Dantas TNC, Júnior HS, Gurgel A. Applications of micelle and microemulsion systems containing aminated surfactants synthesized from ricinoleic acid as carbon-steel corrosion inhibitors. Colloids and Surfaces A: Physicochemical and Engineering Aspects. 2009;**340**:199–207. DOI: 10.1016/j.colsurfa.2009.03.031

[47] Estes TN. Rubber and vinyl protectant. 5,661,208 USA. August 26, 1997. Grant

[48] Porter CA. Weatherable coating and stain system for thermoset or thermoplastic composite surfaces. US 6,201,057 B1 USA. March 13, 2001. Grant

[49] Parvinzadeh M, Hajiraissi R. Macro and microemulsion silicone softeners on polyester fibers: Evaluation of different physical properties. Journal of Surfactants and Detergents. 2008;**11**(4):269–273. DOI 10.1007/s11743-008-1081-7

[50] Ge Y, Gao T, Wang C, Shah ZH, Lu R, Zhang S. Highly efficient silica coated CuNi bimetallic nanocatalyst from reverse microemulsion. Journal of Colloid and Interface Science. 2017;**491**:123–132. DOI: 10.1016/j.jcis.2016.12.043

[51] Małecka MA, Kępiński L. Solid state reactions in highly dispersedpraseodymium oxide–SiO_2system. Journal of Alloys and Compounds. 2007;**430**(1-2):282–291. DOI: 10.1016/j.jallcom.2006.05.011

[52] Kurnatowska M, Kepinski L, Mista W. Structure evolution of nanocrystalline $Ce_{1-x}Pd_xO_{2-y}$ mixed oxide in oxidizing and reducing atmosphere: Reduction-induced activity in low-temperature CO oxidation. Applied Catalysis B: Environmental. 2012;**117–118**:135–147. DOI: 10.1016/j.apcatb.2011.12.034

[53] Kurnatowska M, Kepinski L. Structure and thermal stability ofnanocrystalline $Ce_{1-x}Rh_xO_{2-y}$ in reducing and oxidizing atmosphere. Materials Reserach Bulletin. 2013;**48**(2):852–862. DOI: 10.1016/j.materresbull.2012.11.076

[54] Destrée C, Nagy JB. Mechanism of formation of inorganic and organic nanoparticles from microemulsions. Advances in Colloid and Interface Science. 2006;**123–126**:353–367. DOI: 10.1016/j.cis.2006.05.022

Section 2

Practical Applications of Microemulsions

Chapter 2

Development of Microemulsion Dermal Products Based on Avocado Oil for Topical Administration

Lacramioara Ochiuz and Manuela Hortolomei

Additional information is available at the end of the chapter

http://dx.doi.org/10.5772/66077

Abstract

The research described in this study aimed at developing microemulsions for dermal application using avocado oil. Due to its composition, avocado oil helps maintaining the barrier function of the skin. It has a nutritional effect on the skin, and it reduces the intensity of the process of skin peeling. Various surfactant:cosurfactant systems were tested in the conducted studies. There were no significant differences between the diagrams generated by Tween 20 and the surfactant:cosurfactant system, Tween 20:PEG400, at a ratio of 1:1. Six formulations were selected from the dilution line 7 of the ternary phase diagrams obtained by using as a surfactant Tween 20 and Tween 20:PEG 400, respectively. The formulations were characterized by determining physicochemical properties specific. In the next phase of study, these six formulations were used as a vehicle for incorporating erythromycin in order to develop erythromicyn incorporated formulations for topical administration. The quality control of microemulsions with erythromycin was performed by evaluating the physical chemical, organoleptic and sensorial properties. Microemulsions were pharmacotechnically characterized by assessing the *in vitro* and *ex vivo* release kinetics of erythromycin.

Keywords: microemulsion, avocado oil, dermal product

1. Introduction

Although microemulsions possess numerous advantages as drug delivery systems, these modern pharmaceutical formulations have a limited applicability in the pharmaceutical and cosmetic field, due to the quality conditions for raw materials imposed by current legislation. Lately, a special interest has been shown by the specialists of the pharmaceutical community in formulation and optimization of microemulsions, as an alternative dosage form to conventional emulsions and to other delivery systems for various administration routes. [1, 2].

Avocado oil is a vegetable oil obtained by cold press extraction from the *Persea gratissima* tree fruit (*Lauraceae fam.*), and containing a high amount of saturated and unsaturated fatty acids and other ingredients such as fat-soluble vitamins, minerals, and oligoelements [3]. Due to these bioactive ingredients, avocado oil plays a role in maintaining the skin's barrier function, manifests a nurturing effect on the dermal tissue, and decreases the exfoliation process.

Acne is a common skin condition that affects about 30% of the world's population, and half of whom are patients aged between 30 and 50 years [4]. Topical antibiotic therapy is an integral part of the therapeutic process of acne. Erythromycin is a macrolide antibiotic used to treat acne for over 20 years [5, 6]. Although some studies have shown the emergence of erythromycin resistance of germs involved in the pathogenesis of acne and the presence of local irritative reactions, this antibiotic is still successfully used by dermatologists as the antibiotic of choice for the local treatment of this dermatosis [7–10]. Erythromycin for topical use is usually formulated as cream. Serdoz et al. studied the incorporation of erythromycin into microemulsions based on mono- and triglycerides of medium chain fatty acids using Tween 80 as a surfactant. The obtained results showed a doubling of ER bioavailability when it was orally administered as emulsion [11].

The aim of this study was to include erythromycin in microemulsion-based preparations obtained with avocado oil and various cosolvent systems designed to increase efficiency and skin tolerance to the antibiotic.

2. Materials and methods

2.1. Materials

Avocado oil (Natural Sourcing LLC, Oxford, England), isopropyl myristate (Sigma-Aldrich, Germany), polyethylene glycol 400 (Sigma-Aldrich, Germany), Tween 20 (Sigma-Aldrich, Germany), erythromycin 99.85% purity (Zhejiang Sanmen Hengkang Pharmaceutical Co., Ltd., China), chromatographically pure acetonitrile (Merck, Germany), disodium phosphate (Merck, Germany) were used. Purified water, double distilled water, and other reagents that conform to the quality requirements of the 10th Romanian Pharmacopoeia were used in preparing microemulsions and in the quantitative analysis.

2.2. Methods

2.2.1. Determination of the ternary phase diagram and preparation of microemulsions

In the studied systems, we used lipophilic phase at 1:1 ratio mixture of avocado oil and isopropyl myristate. The surfactant and cosurfactant of the systems were consequently Tween 20 and also a

1:1 ratio mixture of Tween 20/PEG 400. All formulations contained purified water as hydrophilic phase. The diagrams were built according to the method described by Fanun [12, 13].

2.2.1.1. Parameters involved in the physicochemical characterization of the microemulsions

Electrical conductivity (σ)—determined with Metrohm 712 conductometer (Herisau, Switzerland) with graphite electrode at room temperature (25 ± 2°C). The conductometric cell was calibrated with standard KCl solution.

Rheological behavior—the dynamic viscosity (η) was determined by rotational rheoviscositimeter Rheolab MC120 (Stuttgart, Germany).

pH measurement—was performed using Thermo Orion pH meter (Thermo Fisher, Florida, USA).

Fourier transform infrared spectroscopy (FT-IR spectroscopy)—was done on a Vertex 70 FT-IR spectrophotometer (Bruker, Germany), through spectra measurement of each sample and respectively of the raw materials involved. The obtained spectra were compared standard spectra presented in the literature. All measurements were performed three times at room temperature (25 ± 2°C), with the results expressed as a mean of three determinations (± SD).

2.2.2. Preparation of microemulsions with avocado oil and erythromycin

Six microemulsion formulations were investigated. They were denoted as follows: three microemulsion formulations of the ternary phase diagram made with Tween 20 (F1—F3 and 3 formulae (F4—F6) of the diagram in which the surfactant:cosurfactant Tween 20:PEG 400 in the ratio of 1:1 was used. The formulations were selected from the regions of microemulsion formation based on the ternary phase diagrams, as shown in **Figures 1** and **2**, respectively. They were used within 24 h after preparation. Microemulsions were prepared by gentle warming and stirring of the oil phase, by adding the surfactant based on the ternary phase diagram and finally by including the aqueous phase, dropwise and with stirring at 200 rpm. Microemulsions were left at room temperature for 24 h, after which they were tested.

2.2.2.1. Quality control of microemulsions with erythromycin and avocado oil

Assessment of physical, chemical, and tactile properties—sensory properties of microemulsions, spreading ability, the emollient effect right after use (softness) and consistency were assessed in 10 volunteer patients, who rated products from 1 to 100 according to their satisfaction level, following international guidelines on testing of cosmetics [14, 15].

Evaluation of transepidermal water loss (TEWL) measurement—the measurements were performed using Tewameter TM 210 (Courage & Khazaka, Germany) on groups of six people, aged between 20 and 35 years. The epidermal water loss was measured immediately after applying the emulsion on the anterior surface of the palm (an area exposed to air, with a low

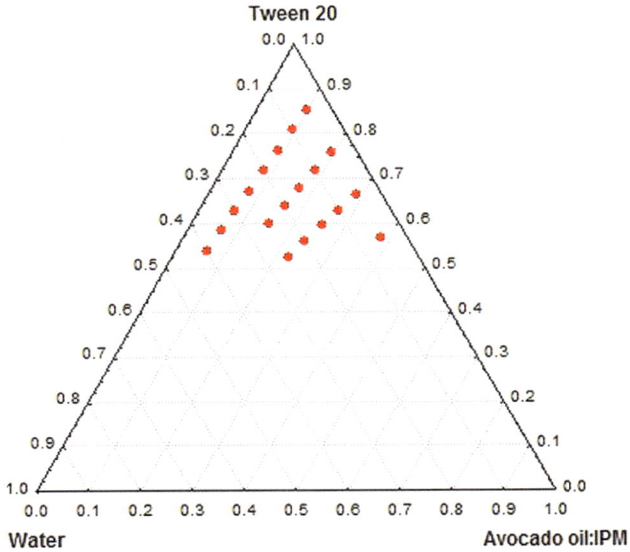

Figure 1. Ternary phase diagrams of the system avocado oil: IPM—Tween 20—water.

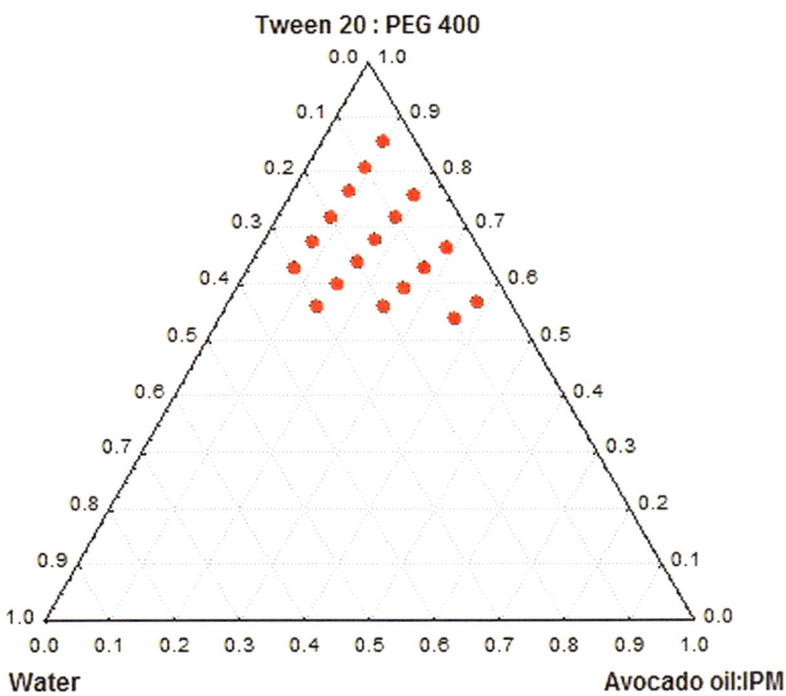

Figure 2. Ternary phase diagrams of the system avocado oil/IPM—Tween 20/PEG 400—water.

density of sebaceous glands). This was considered the starting point of measurements, and it was followed by successive measurements at 30-min intervals for 2 hours [16]. The experiment was carried out for three consecutive days, and the results were expressed as the mean of these three measurements.

Measurement of erythromycin (ER) content in avocado oil-based microemulsions—the measurement of ER content in microemulsions was made by using HPLC method developed and validated in-house [17]. Working method: 0.50 g of sample (F1, F2, F3, F4, F5, And F6) is dissolved separately in 25 mL of the mobile phase, dispersed at warm temperatures for about 5 min by magnetic stirring, and diluted to 10 mL with water. The resulting solution is centrifuged and filtered through a 0.45 μm filter and chromatographed.

2.2.3. *Evaluation of in vitro and ex vivo release of erythromycin in avocado oil-based microemulsions*

An *in vitro* dissolution test was carried out in a Franz cell with a diameter of 2.5 cm and the volume of the acceptor compartment of 15 mL, based on the following protocol: dissolution medium: phosphate buffer pH 7.4; mass of the sample: 0.5 g was processed for each studied formulation; cellulose membrane with a pore size of Ø = 45 μm (Millipore, Merck Germany); temperature: 37 ± 0.2°C; collection interval: 30 min, 1 h, 2 h, 3 h, 4 h, 5 h, and 6 h; an environment volume of 0.5 mL was harvested at each interval, and it was replaced with fresh medium; speed: 100 rpm.

An *ex vivo* dissolution test was carried out under similar conditions and the abdominal skin of female Wistar rats, weighing 200–250 g, was used as biological membrane. After harvesting, hair and adipose tissue were removed from the fraction of the skin and put into a buffer system pH 7.4 for 24 h prior to testing. The collected samples were chromatographed, and erythromycin spectrum was recorded at a wavelength $\lambda = 200$ nm.

2.2.4. *Measurement of the permeability coefficient of erythromycin in avocado oil-based microemulsions*

The permeation rate of the substance in the stationary phase was measured from the slope, and the intercept was measured on the abscissa of the linear portion of the curves of the cumulative amount of drug substance that permeated with respect to time. The permeability coefficient was calculated according to Eq. (1):

$$K_p = J/C \times A \qquad (1)$$

where:

K_p—permeability coefficient (cm/h); J—permeation rate of the drug substance or stationary phase flow (μg/h); C—concentration in the donor compartment (μg/mL); A—contact surface area (cm^2) [18].

3. Results and discussion

Figures 1 and **2** display the ternary phase diagrams of the systems obtained with Tween 20 and Tween 20/PEG 400 (1:1), respectively. The analysis of these diagrams showed no significant differences of the microemulsion zones.

3.1. Physicochemical characterization of microemulsions

Stability assessment was done on six microemulsion formulations selected from the ternary phase diagram built with Tween 20 (1A, 1B, and 1C) and three formulations (2A, 2B, and 2C) selected from the diagram with Tween 20/PEG 400 as surfactant/cosurfactant (1:1 ratio). Formulations were selected on the dilution line 7 of the ternary phase diagrams and contained 5–15% lipophilic phase, with the results shown in **Table 1**.

Formulation	Concentration of hydrophilic phase (% w/w)	Electrical conductivity σ (mS/cm) (± SD)	Dynamic viscosity η (mPa·s) (± SD)	pH (± SD)
1A	5.00	11.98 (±1.21)	40.25 (±0.87)	5.02 (±0.84)
1B	10.00	13.86 (±1.08)	36.11 (±2.01)	5.19 (±0.54)
1C	15.00	17.51 (±0.94)	33.32 (±1.58)	5.21 (±0.60)
2A	5.00	20.50 (±1.42)	58.50 (±1.74)	5.80 (±0.63)
2B	10.00	25.35 (±0.91)	55.01 (±1.01)	5.11 (±0.41)
2C	15.00	29.39 (±1.92)	53.32 (±1.81)	5.35 (±0.31)

Table 1. Physicochemical parameters of microemulsions with avocado oil.

The electrical conductivity of the analyzed microemulsions increases proportionally with the concentration of the hydrophilic phase of the system. The viscosity values indicate a newtonian behavior and a pH value close to the physiological one.

The FTIR spectra of the samples prepared by the two methods are shown in **Figures 3a** and **4a**.

The FTIR spectra recorded for the raw excipients used for the preparation of microemulsions compiled with the reference spectra were presented in the literature [19]. Essentially, the FTIR spectra recorded for the microemulsion samples appear as a combination of the characteristic spectrum of avocado and that of the other three oily excipients. Regarding the resemblance with the spectra recorded for avocado oil, some spectral differences are obvious. First, a very intense and broad band is observed in the spectral range 3100–3700 cm^{-1}, which is the normal range of adsorption for aliphatic hydroxyl group, and it is associated with the water content of the sample. Indeed, this band becomes more intense as the amount of water used in the formulation increases. Secondly, few differences can be noticed in the spectral range 2750–3050 cm^{-1} also in the case of formulations encoded 1A, 1B, and 1C (shown in **Figure 3b**) and for those encoded as 2A, 2B, and 2C (shown in **Figure 4b**). In this area, it is notable the fact that the band at 3009 cm^{-1} assigned to C-H stretching vibration of the *cis*-double bond disappears, while a decrease of the intensity of the band at 2852 cm^{-1} with the formation of a new band at 2874 cm^{-1}. The band at 2852 cm^{-1} is usually assigned to symmetric and asymmetric stretching vibration of the aliphatic$-CH_2$ group, and that from 2874 cm^{-1} corresponds to symmetric and asymmetric stretching vibration of the aliphatic$-CH_3$ group. These observations are supported also by the spectral changes in the region C=O region (1746 cm^{-1}) as it can be seen in **Figures 3c** and **4c**, respectively. Here, the study shows a shift of this band toward lower wavenumbers (1733 cm^{-1}). This may happen due to the formation of saturated aldehyde functional groups or other secondary products during the heating process. These results are in accordance with observations made by other authors on the FTIR study of edible oils [20]. Finally, a new band is observed at 1643 cm^{-1}. This band is associated with the bending vibrations of the covalent bond in water molecules. Its intensity increases with the amount of water used in the formulation (see **Figures 3c** and **4c**).

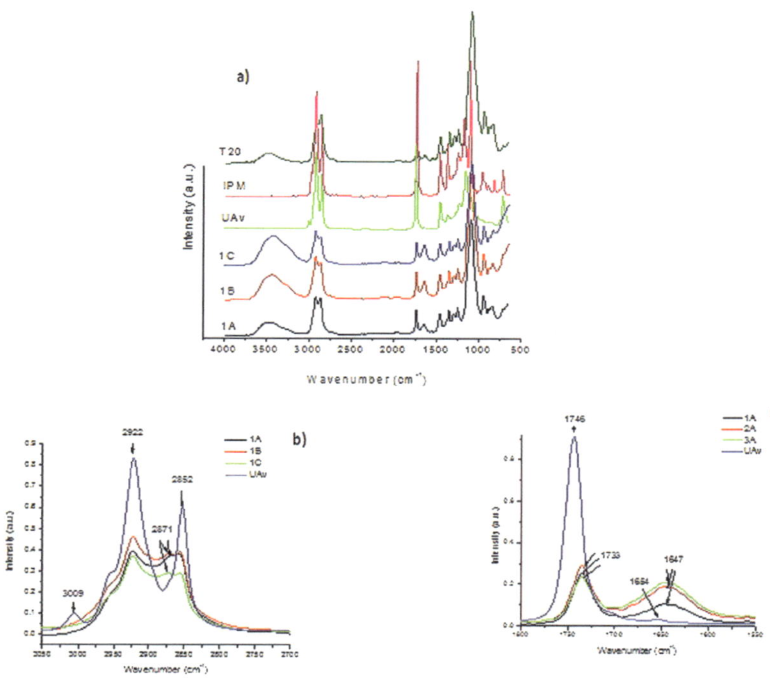

Figure 3. (a) FTIR spectra recorded for microemulsion samples 1A, 1B, 1C, and raw materials avocado oil (UAV), isopropyl miristate (IPM), Tween 20 (T20); (b) FTIR spectra recorded for samples 1A, 1B, 1C, and avocado oil in the 2700–3050 cm^{-1} spectral range; (c) FTIR spectra recorded for samples 1A, 1B, 1C, and avocado oil in the 1550–1800 cm^{-1} spectral range.

In the next phase of the study, these six formulations of avocado oil-based microemulsions were used as a vehicle for incorporating erythromycin in order to develop dermal preparations as microemulsions with erythromycin for topical administration.

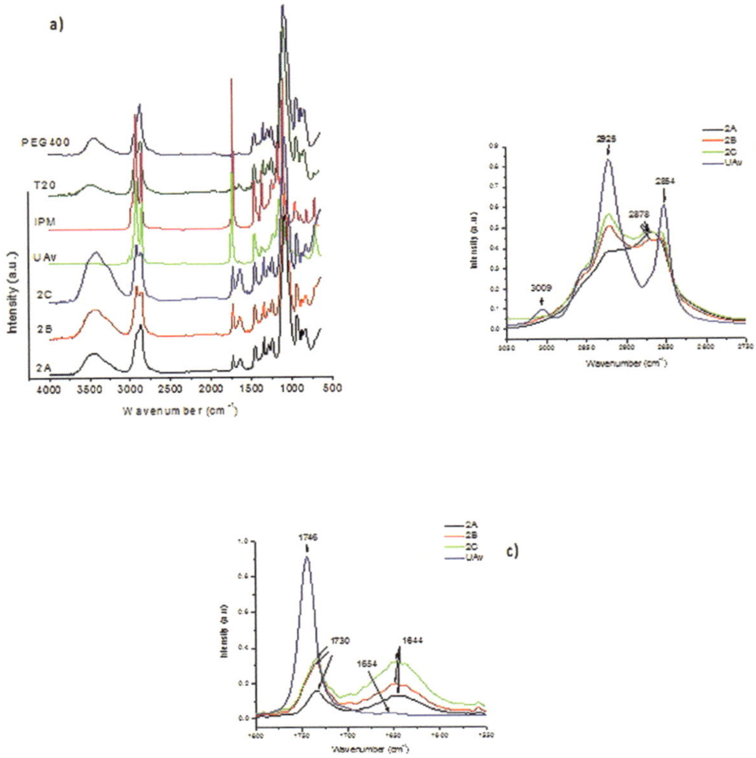

Figure 4. (a) FTIR Spectra recorded for microemulsion samples 2A, 2B, 2C, and raw materials avocado oil (UAV), isopropyl miristate (IPM), Tween 20 (T20); (b) FTIR spectra recorded for samples 2A, 2B, 2C, and avocado oil in the 2700–3050 cm^{-1} spectral range; (c) FTIR spectra recorded for samples 2A, 2B, 2C, and avocado oil in the 1550–1800 cm^{-1} spectral range.

The analysis of sensory properties showed that microemulsions prepared with a mixture Tween 20:PEG 400 as a surfactant had the best moisturizing and spreading properties (**Figure 5**).

This feature can be attributed to the presence of PEG 400 in the formulation, as it has a moisturizing effect on the skin. Although all microemulsion formulations were evaluated with a maximum score of 100 for the aspect, F1 and F3 formulations with the lowest percentage of the aqueous phase had the lowest score for spreading, as a greasy feeling was reported.

The TEWL measurement had the best values for the series of microemulsions prepared with the surfactant system Tween 20:PEG 400, as shown in **Figure 6**.

In order to assess the ER content of the studied microemulsion formulations, we compared the results with data from the monograph *Emulsiones* and *Unguenta* in the 10th Romanian Pharmacopoeia that admits a deviation of ±3% from the declared value for products containing 0.5% and more than 0.5% of active ingredient. The results shown in **Table 2** suggest a uniform distribution of ER in the studied microemulsions, with a content of active ingredient ranging from 97.24 to 103.11%. These values conform to the standards of the 10th Romanian Pharmacopoeia [21].

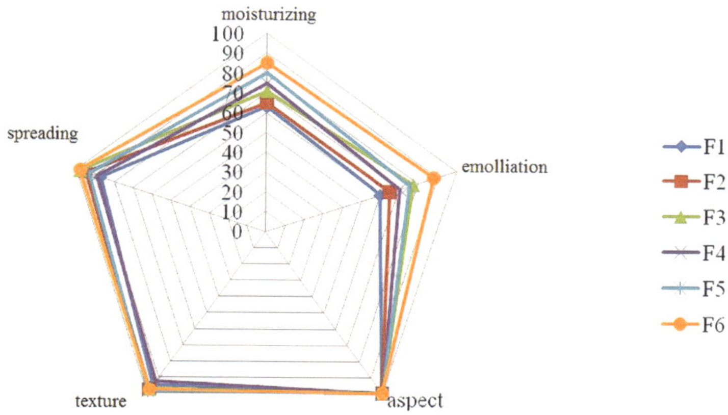

Figure 5. Representation of the sensory properties of avocado oil-based microemulsions with erythromycin.

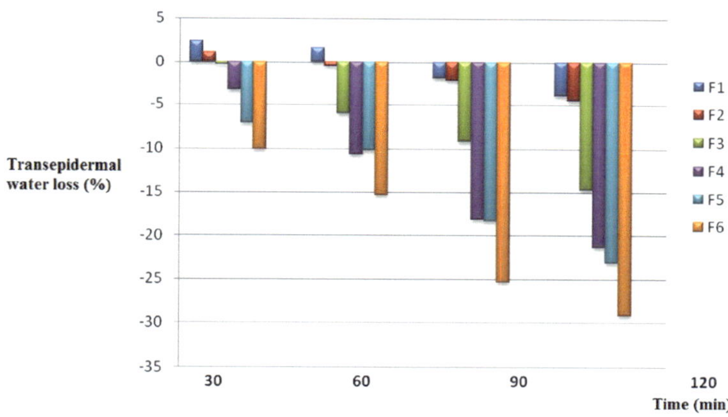

Figure 6. Representation of the TEWL measurement of avocado oil-based microemulsions with erythromycin.

Formulation	The ER content (%)
F1	101.13 ± 0.54
F2	99.25 ± 0.71
F3	97.24 ± 1.04
F4	99.85 ± 0.96
F5	103.11 ± 0.46
F6	100.53 ± 0.87

Table 2. The content of the active ingredient in avocado oil-based microemulsions with erythromycin.

The results of the *in vitro* and *ex vivo* ER release tests in avocado oil-based microemulsions are shown in **Figures 7** and **8**.

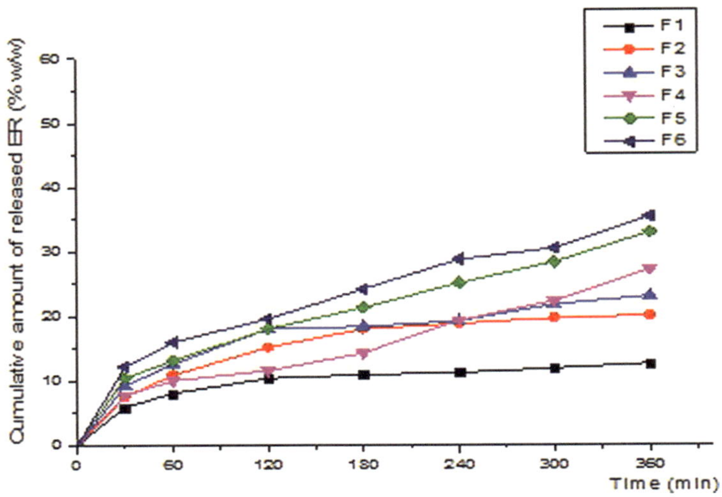

Figure 7. Cumulative kinetic profile of *in vitro* release of ER in avocado oil-based microemulsions.

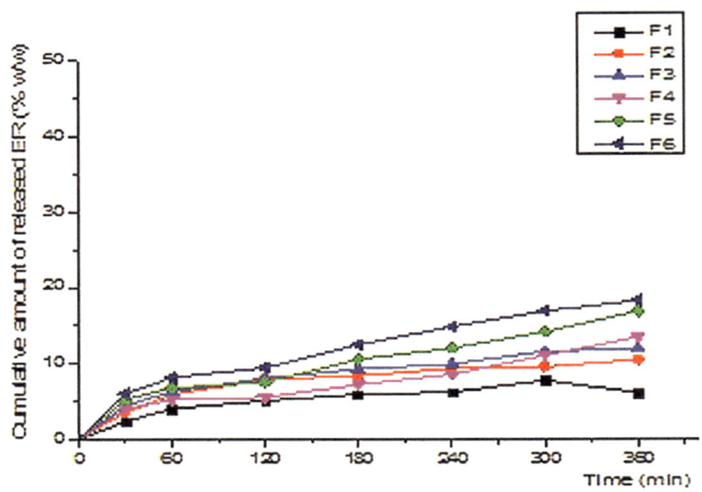

Figure 8. Cumulative kinetic profile of *ex vivo* release of ER in avocado oil-based microemulsions.

The results of the *in vitro* release through the synthetic membrane and *ex vivo* release through the biological membrane showed an unsatisfactory release of ER during the 6 hours of the study. Cumulative analysis of data suggests that F1-F3 formulations prepared with the Tween 20 surfactant had a slightly accelerated release rate within the first 2 hours of testing, after which the concentration of ER remains at a plateau level. We have found that F3 and F6 formulations that had the highest water content showed the highest percentage of release both in testing through biological membrane (F3 = 11.90%, and F6 = 18.18%) and in testing through synthetic membrane (F3 = 22.90 and 35.18%, respectively). F4-F6 formulations prepared with the cosurfactant system Tween 20:PEG 400 showed the highest percentage of ER release, ranging from 26.98 to 35.18% for the *in vitro* test and from 13.45 to 18.18% for the *ex vivo* test. This low percentage of release can be attributed to possible interactions between erythromycin and certain substances found in the microemulsion formulations, to ER insolubility in the dissolution medium, and to the large size of the erythromycin molecule.

Formulations prepared with the cosolvent system Tween 20:PEG 400 had the best coefficient of penetration, due to the presence of PEG 400 in the formulation (**Table 3**).

Formulation	**Permeation parameters**	
	J_{ss} **(μg/cm²/h)**	K_p **× 10⁻⁶ (cm/h)**
	Synthetic membrane (*in vitro*)	
F1	30.4559 ± 7.2197	7115.86
F2	30.3371 ± 7.5029	7561.58
F3	25.0658 ± 7.7158	5424.43
F4	30.3625 ± 7.7720	7882.27
F5	28.2829 ± 7.7888	6926.99
F6	26.9587 ± 7.4907	6156.37
	Biological membrane (*ex vivo*)	
F1	25.0234 ± 7.1224	5181.90
F2	26.8229 ± 7.3379	5922.48
F3	25.3714 ± 7.3734	5466.80
F4	26.6616 ± 7.1809	5839.17
F5	25.0998 ± 7.2846	5387.38
F6	25.2441 ± 7.5083	5465.27

Table 3. Parameters specific to ER permeation in avocado oil-based microemulsions.

4. Conclusions

Microemulsions prepared with the surfactant system Tween 20:PEG 400 had the best sensory properties and the best transepidermal water loss measurement. ER was evenly distributed in the studied microemulsions that had a content of active ingredient ranging from 97.24 to 103.11%. The results of the *in vitro* release tests through synthetic membrane and the results of the *ex vivo* release tests through biological membrane showed an ER release of less than 40%. F4–F6 formulations prepared with the cosurfactant system Tween 20:PEG 400 had the highest percentage of ER release, ranging from 26.98 to 35.18% for the *in vitro* test and from 13.45 to 18.18% for the *ex vivo* test. The permeation degree of erythromycin was influenced by the presence of PEG 400, which led to a significant increase in this parameter.

Author details

Lacramioara Ochiuz[1]* and Manuela Hortolomei[2]

*Address all correspondence to: ochiuzd@yahoo.com

1 Department of Pharmaceutical Technology, Faculty of Pharmacy, University of Medicine and Pharmacy "Grigore T. Popa", Iasi, Romania

2 Ester Farm LTD, Iasi, Romania

References

[1] Bagwe RP, Kanicky JR, Palla BJ, Patanjali PK, Shah DO. Improved drug delivery using microemulsions: rationale, recent progress, and new horizons. Crit Rev Ther Drug Carrier Syst. 2001;**18**:77–140. DOI: 10.1615/CritRevTherDrugCarrierSyst.v18.i1.20

[2] Cui Y, Li L, Gu J, Zhang T, Zhang L. Investigation of microemulsions system for transdermal delivery of ligustrazine phosphate. African J Pharm Pharmacol. 2011;**5**(14):1674–1681. DOI: 10.5897/AJPP11.138

[3] Le Poole HA. Natural oils and fats multifunctional ingredients for skin care. Cosmetics & Toiletries Manufacture Worldwide. 2005;47–56.

[4] Leyden JJ, Del Rosso JQ, Webster GF. Clinical considerations in the treatment of acne vulgaris and other inflammatory skin disorders: focus on antibiotic resistance. Cutis. 2007;**79**:9–25. DOI: 10.1016/j.det.2008.07.008

[5] Eady EA, Gloor M, Leyyden JJ. Propionibacterium acnes resistance: a worldwide problem. Dermatology. 2003;**206**:54–56. DOI: 10.1159/000067822

[6] Eady EA, Ross JI, Cove JH. Multiple mechanisms of erythromycin resistance. J Antimicrob Chemother. 1990;**26**:461–471. DOI: 10.1093/jac/26.4.461

[7] Rosen T, Waisman M. Topically administered clindamycin in the treatment of acne vulgaris and other dermatologic disorders. Pharmacotherapy. 1981;**1**:201–205. DOI: 10.1002/j.1875-9114.1981.tb02541.x

[8] Del Rosso JQ. Emerging topical antimicrobial options for mild-to-moderate acne: a review of the clinical evidence. J Drugs Dermatol. 2008;**7**:2–7.

[9] Del Rosso JQ, Leyden JJ. Status report on antibiotic resistance: implications for the dermatologist. Dermatol Clin. 2007;**25**:127–132.

[10] Cooper AJ. Systematic review of Propionibacterium acnes resistance to systemic antibiotics. Med J Austral. 1998;**169**:259–261.

[11] Serdoz F, Voinovich D, Perissutti B, et al. Development and pharmacokinetic evaluation of erythromycin lipidic formulations for oral administration in rainbow trout (Oncorhynchus mykiss). Eur J Pharm Biophar. 2011;**78**:401–407.

[12] Fanun M. Formulation and characterization of microemulsions based on non-ionic surfactants and peppermint oil. J Colloid Interf Sci. 2010;**343**:496–503.

[13] Fanun M. Properties of microemulsions based on mixed nonionic surfactants and mixed oils. J Mol Liq. 2009;**150**:25–32.

[14] Colipa Guidelines. Efficacy Evaluation of Cosmetic Products [Internet]. May 2008. Available from: https://www.cosmeticseurope.eu/publications-cosmetics-europe- [Accessed: 2015-09-03]

[15] Colipa Guidelines. Test Guidelines for the Assessment of Human Skin Tolerance of Potentially Iritant Cosmetic Ingredients [Internet]. 1997. Available from: https://www.cosmeticseurope.eu/publications-cosmetics-europe- [Accessed: 2015-09-03]

[16] Carey JM, Moran B, Shuster F, et al. Simpler tools for customising sensorial properties. Personal Care. 2008;**9**:89–94.

[17] Hortolomei M, Ochiuz L, Popovici I, Timofte D, Petrescu CD, Ghiciuc CM. Development and validation of the high performance liquid chromatography method for the quantitative determination of erythromucin in dermo-preparations. Rev Med Chir Soc Med Nat lasi. 2015;**119**(4):1174–1179.

[18] Bronaugh RL, Kraeling MEK, Yourick JJ. Determination of percutaneous absorption by in vitro techniques. In: Bronaugh RL, Maibach HI, editors. Percutaneous Absorption. Drugs-Cosmetics-Mechanisms-Methodology. 4th ed. Boca Raton: Taylor & Francis Group; 2005. pp. 265–269.

[19] Dibern HW, Müller RM, Wirbitzki E, editors. UV and IR Spectra: Pharmaceutical Substances (UV and IR) and Pharmaceutical and Cosmetic Excipients (IR). Aulendorf, Germany: ECV, Editio Canto Verlag; 2002.

[20] Vlachos N, Skopelitis Y, Psaroudaki M, Konstantinidou V, Chatzilazarou A, Tegou E. Applications of Fourier transform-infrared spectroscopy to edible oils. Analytica Chimica Acta. 2006;**573–574**:459–465.

[21] The 10th Romanian Pharmacopeea. Bucuresti: Ed. Medicala; 1993. 951p.

Chapter 3

Microemulsion Applications in Carbonate Reservoir Stimulation

Christopher N. Fredd, Mark L. Hoefner and
H. Scott Fogler

Additional information is available at the end of the chapter

http://dx.doi.org/10.5772/65973

Abstract

Carbonate reservoir stimulation involves the injection of reactive fluids, most commonly hydrochloric acid (HCl), into the porous media to enhance the permeability and increase hydrocarbon production. This process results in the formation of highly conductive flow channels, or wormholes, and relies on the deep penetration of reactive fluids into the formation to maximize stimulation success. However, the rapid rate of reaction of HCl with the carbonate rock often limits the depth of live acid penetration. The reaction is mass transfer limited under typical reservoir conditions. As a result, the acid diffusion and convection rates significantly influence the success of the treatments. Microemulsions prepared with HCl as the dispersed phase offer a solution to significantly reduce the effective diffusivity and, hence, increase the depth of stimulation. This chapter presents the results of laboratory studies of carbonate dissolutions using acid microemulsions and highlights case histories of industry applications using macroemulsions for carbonate reservoir stimulation.

Keywords: microemulsion, emulsified acid, reservoir stimulation, carbonate acidizing, fracture acidizing, retarded acid, diffusion coefficient, rotating disk, wormhole, Damköhler number

1. Introduction

Microemulsion technologies have various applications in the oil and gas industry including improved fluid recovery and relative permeability enhancement in drilling and stimulation applications [1–4], displacement and cleanup of oil-based muds [5], and enhanced oil recovery

[6, 7]. Microemulsion properties such as ultralow interfacial tension, large interfacial area, and the ability to solubilize both aqueous and oil-soluble compounds leverage these applications. When used in reservoir stimulation treatments to improve fluid recovery and relative permeability effects, microemulsions are generally used as an additive to the main treating fluid, with a low concentration of a microemulsion being dispersed in the main oil- or aqueous-based fluid. For example, aqueous hydrochloric acid (HCl) treating fluids containing microemulsion additives have been used to enhance stimulation effectiveness by reducing the surface tension and fluid leak-off rates into the porous media during carbonate acidizing treatments [4].

Microemulsions have also been investigated as the main treating fluid for use in carbonate acidizing treatments. For this application, acid microemulsions are formulated with HCl as the dispersed phase, and the microemulsion property of most interest is the well-known ability to reduce the mobility of the dispersed phase. Acid microemulsions were initially studied to improve the success of carbonate acidizing for Danian chalk reservoirs in the North Sea [8]. Later studies evaluated microemulsion properties including the effective diffusivity of the dispersed acid phase and their overall applicability for carbonate acidizing [9, 10]. The use of acid microemulsions as the main treating fluid during carbonate acidizing is the focus of this chapter.

Acidizing is a reservoir stimulation technique commonly used to increase the near-wellbore flow capacity of petroleum production and injection wells. In carbonate reservoirs (comprised of limestone and/or dolomite), the technique involves the injection of acid, typically hydrochloric acid (HCl), into the reservoir to dissolve the formation rock and materials that plug the pore space. The dissolution removes resistances to flow and creates highly conductive flow channels, thereby allowing oil or gas to flow more readily. The key parameter governing channel formation is the Damköhler number which is the ratio of the rate of reaction of the acid with the porous medium to the rate of convective transport of the acid through the rock [9–11]. When the Damköhler number is high, such as would be the case for low acid flow rates or high acid reaction rates, the acid will not penetrate very deep into the rock, resulting in face dissolution and little if any penetration and dissolution into the formation. For the case of low Damköhler numbers, such as high acid flow rates or low rates of acid reaction, the acid dissolves the porous medium relatively uniformly at the expense of large volumes of acid. At intermediate Damköhler numbers, highly conductive flow channels or wormholes are formed, which can efficiently penetrate deep into the formation. This dissolution phenomenon can be easily characterized by laboratory core flood experiments in which HCl is injected into carbonate porous media.

The success of carbonate acidizing treatments relies on the formation of wormholes that penetrate deep into the formation. However, the treatments are often constrained by the very rapid rate of reaction between HCl and carbonate rock, which results in limited depth of live acid penetration and ineffective flow capacity of the acid-dissolution channels [9–17]. The rapid HCl-carbonate rock reaction is limited by the rate of mass transfer of acid to the rock surface (diffusion limited) under common reservoir conditions [18, 19]. This means that the dissolution process can be controlled by changing the rate of convection and/or the rate of acid diffusion. In treatment cases where the injection rate cannot be increased owing to pressure drop considerations, the effectiveness of the treatment can be controlled based on the Damköhler number by reducing the acid reaction rate. Because the reaction is mass transfer limited, this reduction can be achieved by reducing the acid diffusivity. This governing parameter provides the fundamental premise behind evaluating acid microemulsion systems to reduce the mobility (and the effective diffusivity) of the dispersed acid phase for use in carbonate acidizing.

This chapter focuses on the application of acid microemulsions for carbonate acidizing and highlights the fundamental research by Hoefner and Fogler [9, 10]. This chapter includes an overview of the fundamentals of carbonate acidizing, laboratory procedures to assess performance for carbonate acidizing, characteristics of the microemulsion systems with HCl as the dispersed phase, and laboratory results for microemulsion impact on carbonate acidizing. Although the research presented in this chapter demonstrate the technical benefits of microemulsions with HCl as the dispersed phase for carbonate acidizing, the authors are not aware of actual acidizing applications of such microemulsion systems in the oil and gas industry. A patent describing an acid internal microemulsion for carbonate acidizing was granted in 1989 and has since lapsed [20]. The most similar industry application is the use of acid-internal "macro"emulsions to reduce the diffusivity and improved acid penetration. The last section of this chapter provides an overview of these macroemulsion case histories and highlights limitations that may be an area of interest for future microemulsion research.

2. Fundamentals of carbonate acidizing

Carbonate acidizing treatments can be categorized as either matrix acidizing or fracture acidizing. In matrix acidizing, the acid is injected into the porous rock matrix to create acid-dissolution channels, commonly referred to as wormholes. In fracture acidizing, the acid is injected at high rates and pressures to create hydraulic fractures in the rock, which then retain flow capacity after closing due to the creation of nonuniform acid-etched channels along the fracture faces. In both cases, the success of carbonate acidizing treatments depends on the flow capacity of the created acid-dissolution channels and the depth to which they penetrate into the formation. High flow capacity and deep penetration are the desired outcome. This section provides an overview of the fundamentals of HCl-carbonate rock reaction, the dissolution phenomenon that leads to the formation of acid-dissolution channels, and their impact on carbonate acidizing treatments.

2.1. Acid-rock reaction

The rate of acid-carbonate rock reaction depends on the mineralogical composition of the rock formation. Carbonate reservoirs are composed of limestone ($CaCO_3$) and/or dolomite [$CaMg(CO_3)_2$], which are completely soluble in acid. The dissolution of limestone by HCl is mass transfer limited at temperatures above 0°C [18], while the dissolution of dolomite is mass transfer limited above 50°C [19]. Thus, at temperatures commonly encountered in petroleum reservoirs, the rate of dissolution is highly dependent on the rate at which the acid is transported to the rock surface by convection and diffusion. The carbonate reaction chemistry is described in more detail elsewhere in the literature [21].

2.2. Dissolution phenomenon

The flow and reaction of fluids in carbonate porous media results in the formation of highly conductive flow channels, referred to as wormholes (**Figure 1**). Wormholes are desirable during matrix acidizing treatments because they are capable of bypassing near-wellbore damage and their conductivity is several orders of magnitude higher than that of the porous medium. Wormholes form because of the rapid and almost complete dissolution of the mineral in HCl and the variability in pore-scale flow rates caused by the natural heterogeneity of the porous medium. The structure of the dissolution channels vary significantly, depending on parameters such as the injection rate, effective diffusion coefficient, and fluid-mineral properties. **Figure 1** illustrates the five main types of dissolution structures:

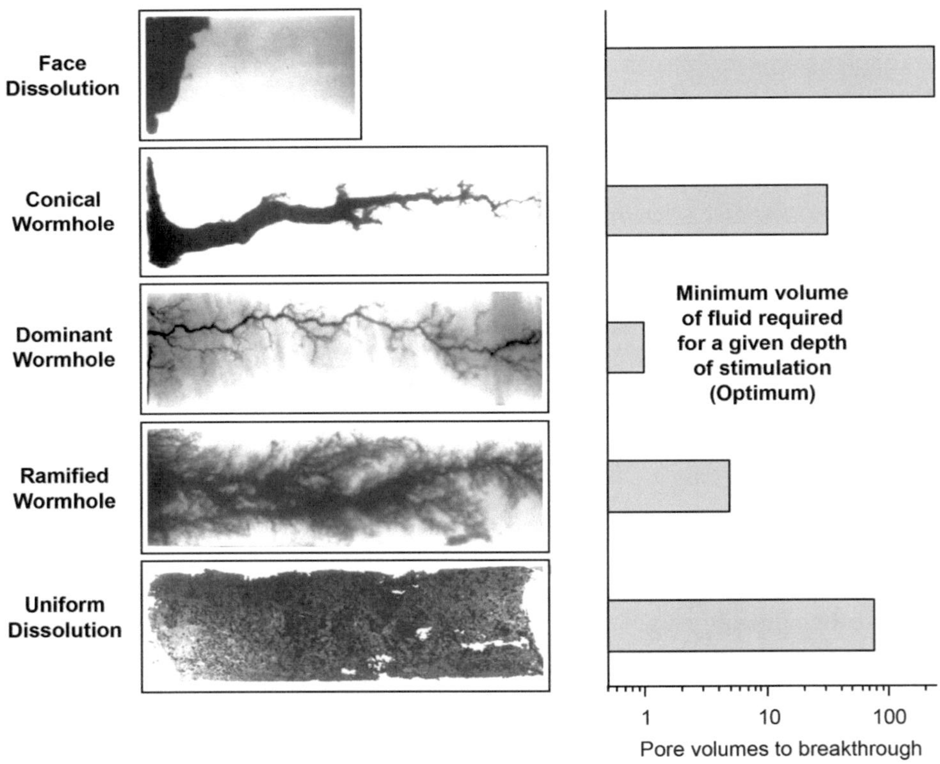

Figure 1. Dissolution structures formed during flow and reaction in carbonate porous media, and corresponding pore volumes of fluid required for breakthrough. (Fluids were injected from left to right during linear coreflood experiments.) The top four images are neutron radiographs of dissolution structures formed during the dissolution of limestone by hydrochloric (HCl) and ethylenediaminetetraacetic acid (EDTA) [11, 22, 23]. The bottom image is a Wood's metal casting of a dissolution structure formed during the dissolution of dolomite by aqueous HCl [10].

1. Face dissolution
2. Conical wormholes
3. Dominant wormholes
4. Ramified wormholes
5. Uniform dissolution

The dissolution structure transforms from face dissolution to uniform dissolution (from top to bottom in **Figure 1**) as the injection rate is increased or the effective diffusion rate is decreased. At low injection rates or high diffusion rates, the reactant is consumed on the inlet flow face of the core, resulting in complete dissolution of the core starting from the inlet flow face. This face dissolution consumes large volumes of reactant and provides negligible depths of live acid penetration. (The required volume of fluid is demonstrated by the bar graph in **Figure 1**, where one pore volume equals the initial porosity times the bulk core volume; and pore volume to breakthrough is the injected pore volume at which the dissolution structure propagates the length of the core.) At slightly higher injection rates or lower diffusion rates, the reactant can penetrate into the porous matrix and enlarge flow channels. However, conical-shaped dissolution channels are formed due to a significant amount of reactant being consumed on the channel walls. At intermediate injection rates, reactant is effectively transported to the tip of the propagating flow channel, where reaction further extends the channel length and creates a dominant wormhole structure. This dominant wormhole structure minimizes the volume of fluid required to achieve a given depth of stimulation and represents the most effective conditions for matrix stimulation. At high injection rates or low diffusion rates, fluid is transported deeper into the porous medium before reaction, resulting in the formation of highly branched or ramified wormholes and an increase in the required volume of fluid. At the extreme of very high injection rates or very low diffusion rates, the reactant is transported to most pores in the medium giving rise to uniform dissolution and inefficient matrix stimulation. Hence, the dominant wormhole structure represents the most effective conditions for matrix stimulation and an optimum injection rate exists for a given fluid-rock system at which a minimum volume of fluid is required to achieve a given depth of wormhole penetration [10, 11, 14–17, 22, 23]. These results demonstrate that the wormhole formation has a significant impact on the success of carbonate stimulation treatments.

The dissolution structure ultimately controls the effectiveness of carbonate acidizing treatment, which has led many investigators to study the dissolution phenomenon. Early investigators recognized the significant influence of mass transfer on wormhole formation in limestone [12, 13]. This influence has served as a basis for many of the theories describing wormhole formation and has led to models of wormhole formation that depend on dimensionless terms such as the Peclet number (ratio of rates of transport by convection to transport by diffusion) [14–16] and the Damköhler number (ratio of the overall rate of dissolution to the rate of transport by convection) [10, 11]. Both dimensionless terms depend on the effective diffusion coefficient and successfully predict wormhole formation for at least a few types of dissolution structures.

The Damköhler number describes the phenomenon of wormhole formation for a wide range of fluid/mineral systems and will be used in this chapter. The Damköhler number, N_{Da}, is defined as the ratio of the overall rate of dissolution to the rate of transport by convection and is given by

$$N_{Da} = \frac{\pi d L \kappa}{q} \tag{1}$$

where q is the flow rate in the wormhole, d and L are the diameter and length of the wormhole, respectively, and κ is the overall dissolution rate constant that depends on the fluid-rock system.

When the overall rate of dissolution is mass-transfer limited, κ is given by

$$\kappa_{mt} = \frac{1.86 D_e^{2/3}}{d} \left(\frac{4q}{\pi L}\right)^{1/3} \tag{2}$$

where D_e is the effective diffusion coefficient, which depends on the fluid system.

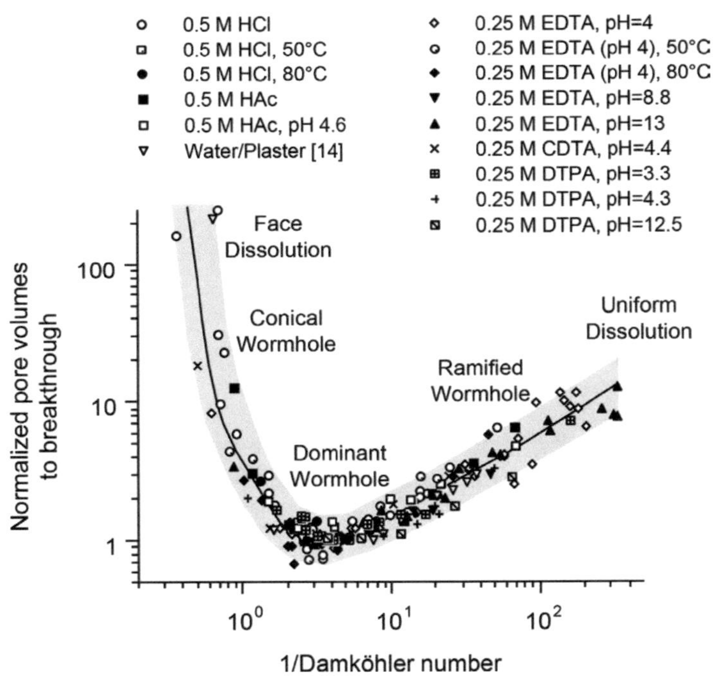

Figure 2. Dependence of the number of pore volumes to breakthrough on the Damköhler number for a wide range of fluid-rock systems (adapted from Ref. [11]).

The Damköhler number captures the balance between convention, diffusion, and reaction that takes place during flow and reaction in porous media and governs the structure of the wormhole channels. This dependency is illustrated in **Figure 2** where the type of dissolution structure and corresponding normalized number of pore volumes to breakthrough is shown as a function of the Damköhler number for a wide range of fluid/rock systems (adapted from Ref. [11]). The data includes aqueous HCl, organic acids [acetic acid (HAc) and formic acid (HFc)], and chelating agents [ethylenediaminetetraacetic acid (EDTA), 1,2-cyclohexane-diaminetetraacetic acid (CDTA), and diethylenetriaminepentaacetic acid (DTPA)], which have varying effective diffusion coefficients and are influenced by reversible or irreversible surface reactions. The results demonstrate the existence of an optimum Damköhler number at which dominant wormhole structures are formed and a minimum volume of fluid is required to achieve a given depth of wormhole penetration. This optimum Damköhler number is observed at 0.29 for several fluid-rock systems and can provide a design basis for optimizing matrix acidizing treatments [11]. A similar dependence on a Damköhler number was demonstrated for acid etching along a fracture face during fracture acidizing [24], and other investigators have highlighted the importance of mass transfer in etching of the fracture face [25, 26]. Hence, both carbonate acidizing approaches are highly dependent on the rate of mass transfer and the effective diffusion coefficient.

2.3. Implications for carbonate acidizing treatments

During matrix acidizing treatments, the low injection rates commonly lead to a Damköhler number higher than 1, where the rapid HCl reaction rate results in face dissolution and an

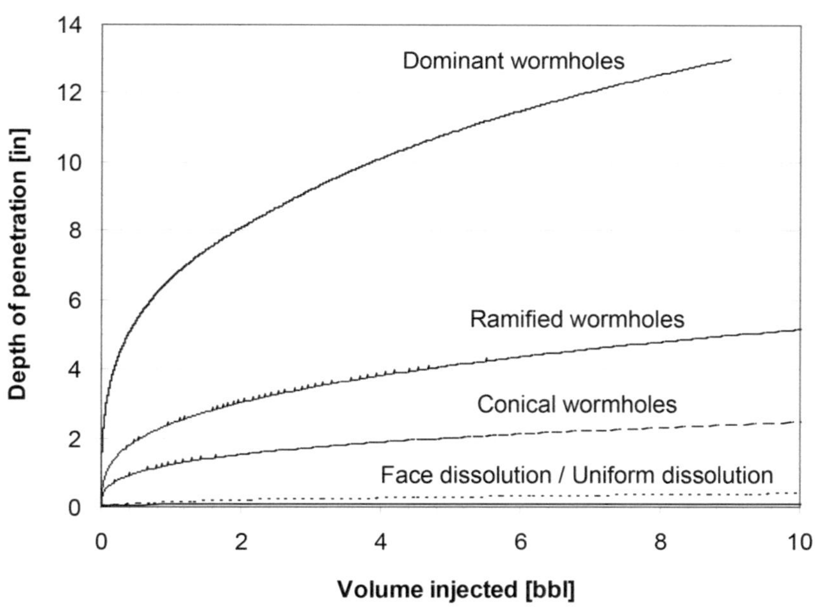

Figure 3. Effects of dissolution structure on the depth of penetration [23]. Copyright 2000, Society of Petroleum Engineers Inc. Reproduced with permission of SPE. Further reproduction prohibited without permission.

ineffective treatment. The impact of the dissolution structure on the effectiveness of matrix acidizing treatments is demonstrated in **Figure 3**, where the depth of penetration into the formation is shown as a function of the volume of HCl injected into limestone at constant Damköhler numbers [23]. When face dissolution occurs, the acid penetrates only a fraction of an inch into the formation. Conical wormholes result in a slight increase to about 2 inches of penetration. Near the optimum Damköhler number, wormhole formation results in effective stimulation as dominant wormholes penetrate over 12 inches into the formation. As the Damköhler number was decreased and the dissolution structure changed to ramified wormholes and uniform dissolution, the depth of penetration decreased. This trend is consistent with experimental and theoretical results reported by several investigators, who report that wormholes typically penetrate a few feet (0.3–2 m) radially into the formation [12, 13]. At these depths, wormholes stimulate the near-wellbore region (where the majority of flow resistance associated with the radial geometry is manifested) and provide an effective matrix acidizing treatment.

Analogous challenges are encountered during fracture acidizing. Danian chalk reservoirs in the North Sea exhibited poor performance after fracture acidizing due to insufficient acid-etched channel flow capacity and failure of the channels to withstand the closure pressure as a result of low rock hardness [8]. Other investigators demonstrated the effects of acid on rock strength in both chalk formations and competent carbonate formations from laboratory studies and recommended minimizing unnecessary dissolution of the fracture face to better maintain rock strength [26]. These examples demonstrate the common challenges of rapid acid reaction rates resulting in ineffective flow capacity of acid-dissolution channels and the need to reduce the rate of dissolution to enable deeper live acid penetration for both matrix and fracture acidizing treatments in a wide range of carbonate formations.

3. Laboratory methods

Several microemulsion systems (described in the following section) were identified for laboratory study and potential application in the oil and gas industry. Effective acid diffusion coefficients were measured for different surfactant and cosurfactant combinations as a function of acid phase volume using a rotating disk technique while in some cases a NMR self-diffusion technique. Trends in the acid diffusion rate as a function of microemulsion composition (and associated structure) were measured to help optimize the system for a particular application, e.g., to optimize acid strength or temperature stability.

The effectiveness of various microemulsion systems and compositions in increasing rock permeability as a function of distance from the wellbore was studied using core flow experiments relating live acid penetration depth to injection volume and conditions (e.g., injection rate and effective diffusion coefficient). Imaging techniques were then used to characterize the structure of the flow channels created in the rock and in turn relate channel structure to live acid diffusion rate and fluid injection rate. Analyses (including numerical network modeling, which is outside the scope of this chapter) were then used to relate the properties of the acid

http://dx.doi.org/10.5772/65973

system, the injection conditions, and the resulting dissolution channel structure to the overall effectiveness of the acidizing process in increasing permeability away from the wellbore.

3.1. Rotating disk technique for measuring diffusion rates

Acid diffusion rates as a function of microemulsion system and composition were studied using a rotating disk system [27, 28]. Polished marble disks ($CaCO_3$) were partially dissolved by rotation in the acid microemulsion. The measurements were carried out under high inert gas overburden pressure to prevent the evolution of gaseous reaction products that would interfere with hydrodynamics at the surface of the disk. The rate of dissolution was measured by analyzing samples for calcium using atomic absorption spectroscopy. For the diffusion-controlled reaction (rapid reaction between HCl and the limestone rock), the rate of acid transport to the surface of the disk is proportional to the disk rotation speed to the ½ power and to the acid diffusion rate to the ⅔ power [27]. For the apparatus used, correction factors that account for finite disk diameter and reactor volume can be neglected.

The dissolution rate of the carbonate disk was monitored by sampling the liquid concentration for calcium. By measuring the dissolution rate at different disk rotational speeds, an effective acid diffusion coefficient for the microemulsion system can be measured. **Figure 4** shows data for a typical set of experiments. For a mass transfer limited reaction, the dissolution rate plotted against the square root of the disk rotation speed will give a straight line passing through the origin. From the slope, we calculate the effective diffusivity, D_e, for the acid microemulsion system.

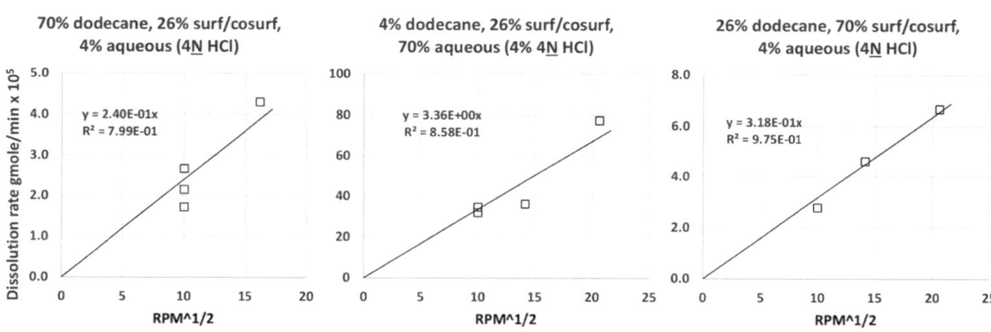

Figure 4. Rotating disk dissolution rate results for three microemulsion formulations.

3.2. Core flow experiments

It is challenging to carry out controlled experiments in actual oil wells. In order to simulate flow and reaction taking place in the rock during acidizing, laboratory "coreflood" experiments are often performed. Core-flooding is a general term and many processes related to petroleum production can be studied using the procedure. Such experiments can be extremely complex, intended to quantitatively represent in situ flow conditions related to hydrocarbon recovery, or they can be simplified to examine one or a few individual effects. In this case, the acid system such as aqueous HCl or microemulsion is injected axially at a constant rate through cylindrical limestone or dolomite cores and the pressure drop along the length of the core is monitored as the dissolution progresses in the porous medium. Similar to the rotating disk experiments, sufficient backpressure was maintained to prevent gaseous reaction products (CO_2) from evolving in the cores. Coreflood experiments were performed with the cores initially water saturated to eliminate any relative permeability effects associated with an initial oil or gas saturation. Thus, the coreflood experiments were designed to examine the dissolution phenomenon in carbonate cores.

Coreflood experiments were performed to test whether lowering the HCl diffusion rate would allow the microemulsion to achieve deep penetrating, dominant wormhole structures for effective matrix stimulation of carbonate rock. Fluid was pumped at a constant rate axially through 1 inch (2.54 cm) diameter rock cores. The pressure drop across the core was monitored as a function of fluid volume injected. The cell designed to hold the core and simulate reservoir conditions is shown in **Figure 5**. Details of the coreflood experimental procedures and apparatus are described elsewhere in the literature [10, 22, 29].

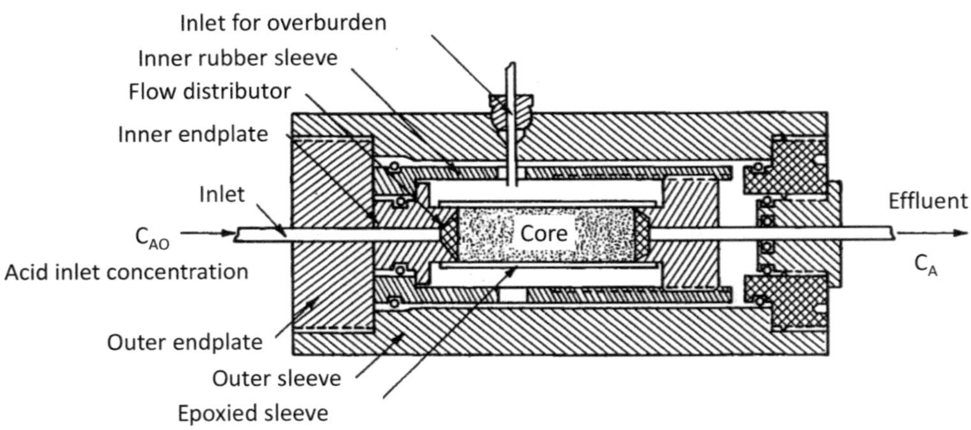

Figure 5. Bi-axial core holder or "Hassler" cell [10]. Copyright© 1988 American Institute of Chemical Engineers.

As the fluid is injected and acid begins to dissolve the rock matrix, the pressure drop associated with the flow begins to decrease, corresponding to an increase in the rock permeability. Results are typically plotted as the ratio of the permeability at any time to the initial permeability as a function of the volume of fluid injected (expressed as porevolumes, where one porevolume equals the initial porosity times the bulk core volume). The pore volume of fluid required to bring about a significant increase in the permeability is a direct measure of the effectiveness of that fluid to stimulate the rock under those treatment conditions. The coreflood experiment provides information relating the effects of parameters such as the fluid injection rate and the acid diffusion rate or reaction rate, as well as properties of the rock such as composition, permeability, and pore size distribution.

Core flow experiments have been conducted over a wide range of core dimensions, rock types (including limestone as well as dolomite), injection rates, and acid diffusion rates for various microemulsion systems.

3.3. Dissolution channel characterization

Imaging techniques were used to evaluate the impact of microemulsion properties (acid diffusion rate) and flow conditions (fluid flow rate) and resulting increase in rock permeability on the structure of the dissolution channels that resulted from the core flow experiments. The earliest technique involved first drying the cores after acidizing, evacuating, and heating in the presence of Wood's metal (a low-melting point mixture of bismuth, lead, tin, and cadmium). The metal was melted and subsequently forced into the channels when the vacuum was released, thus forming a cast of the dissolution channels. The remaining core material was then dissolved in HCl, leaving behind the metal casting of the channel structure. An example is shown in **Figure 1** (bottom image).

Later, a technique involving neutron radiography was used to image the Wood's metal-filled dissolution structures, thereby eliminating the need to dissolve the rock and potential damage to the fine details of the wormhole casting. This technique is ideally suited for imaging structures within consolidated porous media because: (1) the matrix is virtually transparent to thermal neutrons and (2) the cadmium-containing Wood's metal is an excellent neutron absorber and thus provides high contrast between the dissolution channels and the consolidated porous matrix [30]. To image the wormholes, the Wood's metal-filled cores were placed in a beam of thermal neutrons for imaging. Neutron radiography is described in more detail elsewhere in the literature [30, 31].

4. Microemulsions

The fundamental premise behind evaluating microemulsion systems for use in carbonate acidizing is that microemulsions have the well-known ability to reduce the mobility of the dispersed phase, in this case hydrochloric acid, without a proportionate increase in the fluid viscosity. This ability is important because the very rapid HCl-carbonate reaction rate is limited by the rate of diffusion of acid to the rock surface ("diffusion-limited"). One could, in theory, decrease the acid diffusion rate by increasing the fluid viscosity, for example with polymer viscosifiers, but the result would be slower fluid flow through the porous media. Slower fluid flow would in-turn offset the effect of slowing the acid diffusion rate by allowing acid more time to diffuse to the rock surface, as governed by the Damköhler number. It should be emphasized that most matrix acidizing treatments are limited in the rate that the fluid can be injected into a well to avoid fracturing the rock (i.e., the injection rate is often already too low for efficient wormhole formation). In addition, the acid diffusion coefficient increases with increasing temperature, making it even more important to control the acid diffusion rate in high temperature reservoirs. Hence, acid microemulsions provide a potential solution for controlling the acid diffusion rates without significantly reducing the flow rates and in high temperature reservoirs. This section reviews the characteristics of acid microemulsion systems for use in carbonate acidizing.

4.1. Microemulsion systems

The microemulsion systems that have been investigated for decreasing acid diffusion rates include one using a cationic surfactant, and one using a nonionic surfactant [32, 33]. The cationic system, comprising dodecane, water phase (actually aqueous HCl), butanol, and cetylpyridinium chloride (CPC), was used to demonstrate the capability of the microemulsion system and to understand the relationship between dissolved channel structure and acid effectiveness in increasing rock permeability.

A cationic surfactant was chosen for increased microemulsion stability at higher temperatures, for decreased surfactant adsorption on calcite at low pH, and for surfactant stability in the presence of reaction products. Phase diagrams for the system are shown in **Figure 6**. The lower right comer represents a constant surfactant/cosurfactant weight ratio of 35:65 (corresponding to 10 mole% surfactant). The large enclosed area indicates a one-phase region. The most important factors in choosing the model system are: (1) it can solubilize a significant volume fraction of concentrated HCI, and (2) the one-phase region is extensive to minimize the possibility of viscous surfactant mesophases forming when the microemulsion contacts water or oil already present in the rock.

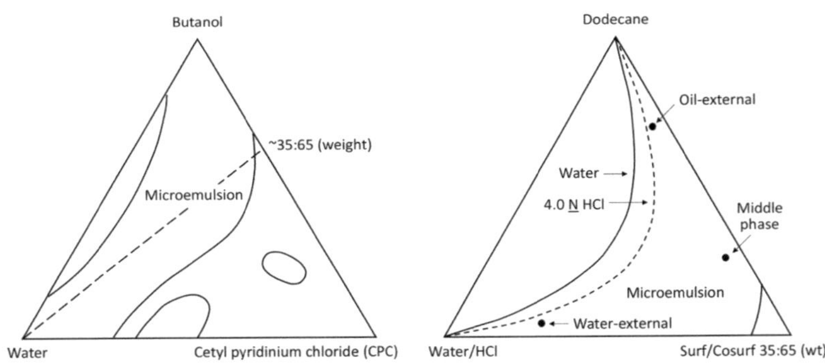

Figure 6. Phase behavior for CPC/dodecane/butanol/water (aqueous hydrochloric acid) system.

The area of the one-phase region near the pure oil corner may represent a reversed micelle structure, while regular micelles (oil-in-water) may exist near the water comer. The center-region may consist of the two micelle types in equilibrium. Reversed micelles are expected to restrict acid mobility (acid diffusivity) to the greatest extent, and so formulations in that region comprised the bulk of the microemulsion studies.

Phase behavior was characterized for the cationic system for cosurfactants ranging from propanol through octanol in order to understand the effects of the cosurfactant chain length. In all cases, the surfactant/cosurfactant ratio was kept at 1:9 molar (corresponding to a 35:65 weight ratio of surfactant to butanol). The butanol system was chosen for further study

http://dx.doi.org/10.5772/65973

because, although lower acid diffusion rates were achieved with pentanol, the isotropic region for butanol was more extensive compared to the longer alcohols.

A nonionic system utilizing mixtures of nonylphenolpolyethoxy alcohol surfactant, naphtha, and aqueous HCl was studied as a potential commercially applicable system. The specific systems involved combinations of surfactants with different numbers of ethoxy head units (Triton brand X-35, N-42, and N-57). The nonionic system had the advantage of significantly lower surfactant cost, no requirement for a cosurfactant, and the ability to solubilize a higher aqueous phase fraction as compared to CPC. A disadvantage of the Triton systems was that the extent of the isotropic region was sensitive to temperature, although by mixing the different surfactants, it was shown that the system could be customized to a broad temperature range and thereby tailored to the specific reservoir temperature of wells to be treated [33].

5. Microemulsions enhancing carbonate reservoir stimulation

The microemulsion systems described in the previous section were evaluated for applications in carbonate acidizing using rotating disk and coreflood experiments.

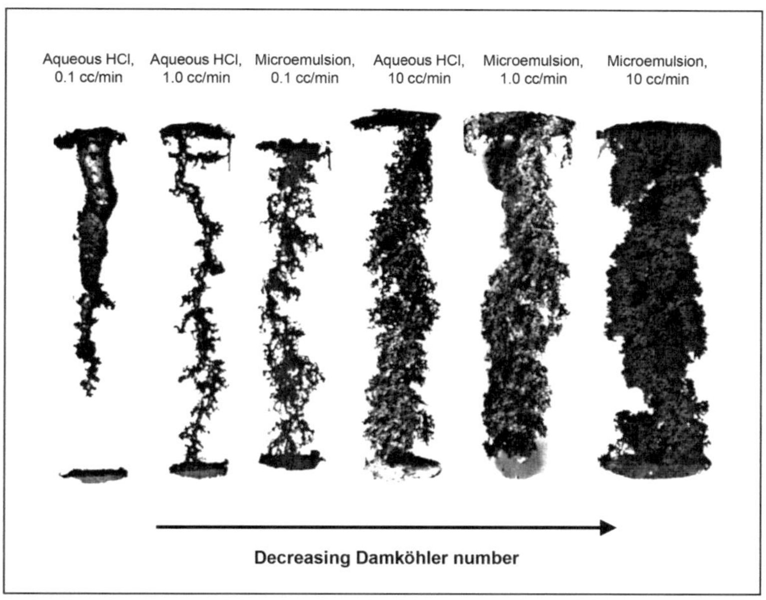

Figure 7. Wormhole structures formed during the flow and reaction of microemulsions and aqueous HCl in limestone, with decreasing Damköhler number from left to right (adapted from Ref. [10]). Copyright© 1988 American Institute of Chemical Engineers.

Rotating disk experiments demonstrated that the dissolution of limestone by acid microemulsion is fully mass transfer limited at 22°C. Data shown in **Figure 4** represent an acid microemulsion formulation near the pure oil corner of the phase diagram that exhibits an acid diffusion coefficient approximately two orders of magnitude lower than for aqueous HCl. The effective acid diffusion coefficient for the microemulsion system is 8×10^{-7} cm^2/s compared to 4×10^{-5} cm^2/s for aqueous HCl [9, 32]. This significant decrease in the acid diffusion coefficient provides a significant reduction in the rate of acid-rock reaction under mass transfer limited conditions. There is also a relatively insignificant decrease in the reaction rate due to a viscosity effect, as the microemulsion has a viscosity of only about 3 cp. For comparison, gelled acids with viscosities greater than 25 cp at 100 s^{-1} are commonly used as "retarded" acid systems, but they reduce the effective diffusivity by only a factor of two compared to aqueous HCl [34]. Hence, the reduced acid mobility provided by the microemulsion system provides far superior reduction of the acid diffusion coefficient than viscous mechanisms.

The transport and reaction of microemulsions in carbonate porous media during coreflood experiments results in the formation of wormholes as demonstrated in **Figure 7**. The images demonstrate that the acid microemulsion created a dominant wormhole structure at 0.1 cm^3/min (third image). At the same injection rate, aqueous HCl resulted in the formation of an inefficient conical dissolution structure (first image). The improved effectiveness of microemulsions is further illustrated in **Figure 8** for the butanol and pentanol formulations. The pore volumes to breakthrough curve for the microemulsions are shifted by about an order of magnitude to lower injection rates relative to aqueous HCl. Hence, the microemulsions are more effective than aqueous HCl at the low injection rates commonly encountered during carbonate matrix acidizing treatments.

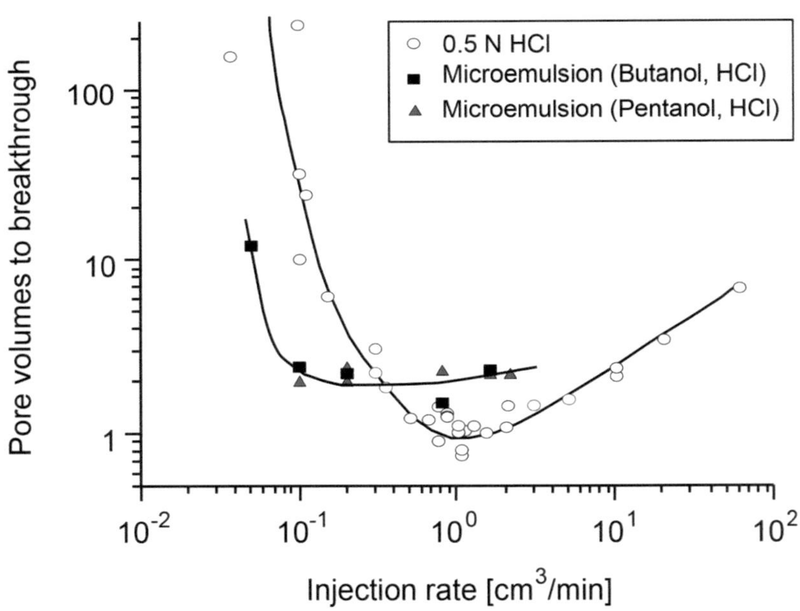

Figure 8. Comparison of stimulation effectiveness for microemulsions and aqueous HCl (data adapted from Refs. [9, 11]).

The improved performance achieved by the acid microemulsion systems at low injection rates is due to the dependence of the dissolution phenomenon on the Damköhler number and the significant reduction in the effective diffusivity of the dispersed acid phase. This dependency is demonstrated in **Figure 9**, where the acid microemulsion systems overlay the performance band observed for the wide range of fluid-rock systems from **Figure 2**, and exhibit the existence of an optimum Damköhler number at which the dissolution process is most effective. Based on these results, acid microemulsions provide useful characteristics for use in carbonate acidizing treatments, particularly at low injection rates and high temperatures where HCl is not effective.

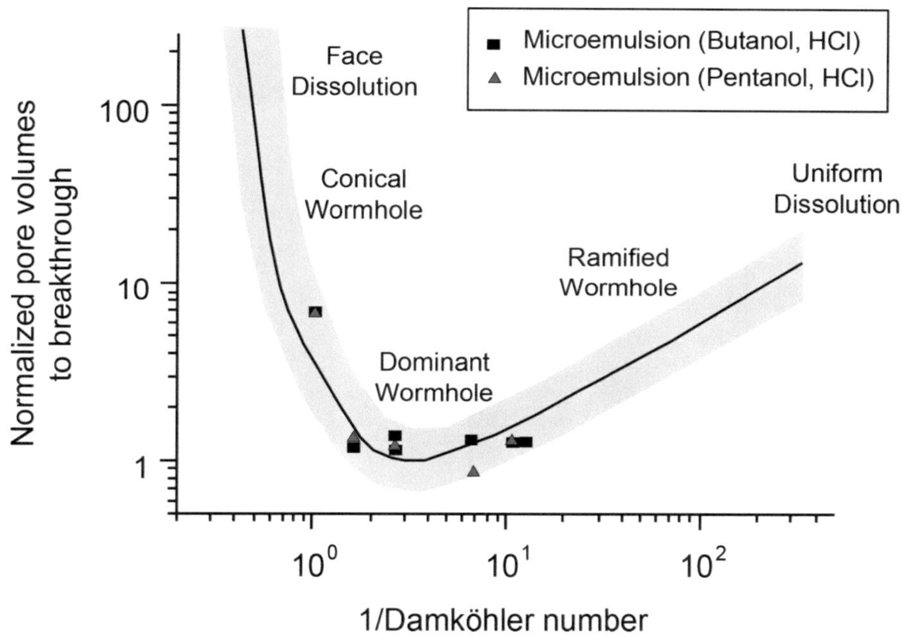

Figure 9. Dependence of the number of pore volumes to breakthrough on the Damköhler number for acid microemulsion systems (data adapted from Ref. [10]). The gray shaded area is the performance band observed for the wide range of fluid-rock systems from **Figure 2**.

6. Industry applications: carbonate acidizing

Microemulsions have numerous properties such as ultralow interfacial tension, large interfacial area, and the ability to solubilize both aqueous and oil-soluble compounds that are valuable for various applications in the oil and gas industry. Carbonate acidizing treatments performed with aqueous HCl as the main treating fluid have contained low concentrations of microemulsion additives to enhance stimulation effectiveness by reducing the surface tension and fluid leak-off rates into the porous media [4]. This chapter highlighted the evaluation of acid microemulsion systems for use as the main treating fluid to enhance dissolution channel formation and the depth of live acid penetration during carbonate acidizing treatments, where the microemulsion property of most interest is the ability to reduce the mobility of the dispersed acid phase.

Although research demonstrates the technical benefits of acid microemulsion systems, the authors are not aware of actual carbonate acidizing applications of such systems in the oil and gas industry. A patent describing an acid internal microemulsion for carbonate acidizing was granted in 1989 and has since lapsed [20]. Since the early 1990s, the industry has commonly used emulsified acids, formulated as acid-internal "macro"emulsions, to reduce the effective acid diffusivity and improve the depth of live acid penetration [25]. Typical macroemulsion systems prepared with HCl as the dispersed phase have effective diffusion coefficients of about $8 \times 10^{-7} cm^2/s$ at 22°C [34], which is consistent with that measured for the microemulsions systems in this research. Hence, the carbonate dissolution phenomenon is similar for acid microemulsions and macroemulsions.

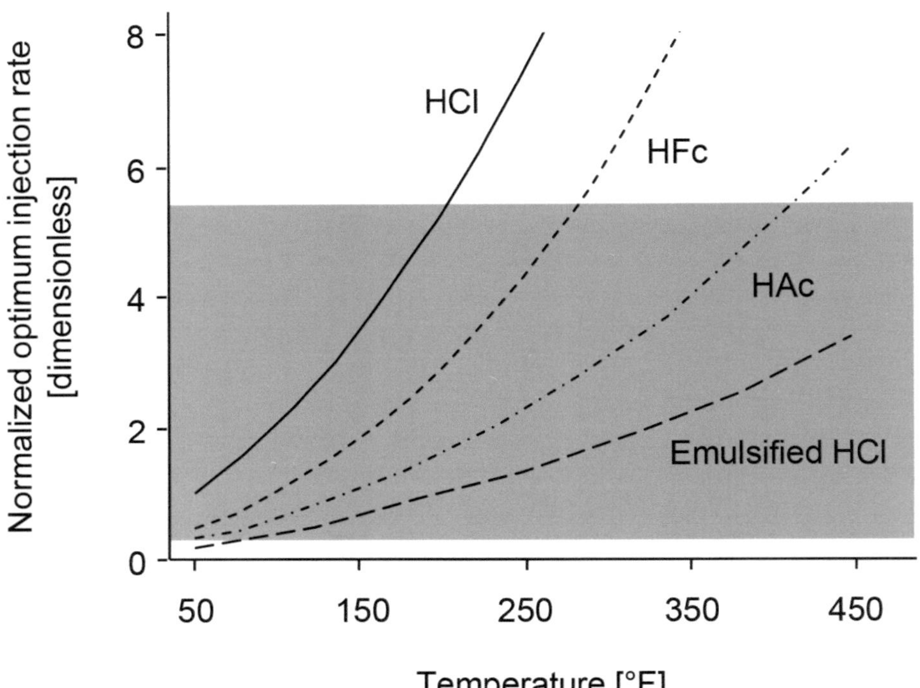

Figure 10. Effect of temperature on the optimum injection rate required to achieve wormhole penetration of 6 inches with various acid systems [23]. (The shaded box represents typical injection rates used in matrix acidizing treatments.) Copyright 2000, Society of Petroleum Engineers Inc. Reproduced with permission of SPE. Further reproduction prohibited without permission.

Macroemulsions are successfully applied in both matrix acidizing and fracture acidizing treatments [25, 35–37]. These case histories demonstrate the use of emulsified acids as the main treating fluid to increase the depth of stimulation at treatments up to about 176°C. The emulsified acids provide improved stimulation effectiveness over a wider range of injection rates and temperatures compared to aqueous acid systems. This wider operating range is illustrated in **Figure 10**, where the optimum injection rate for the creation of dominate wormhole structures is plotted as a function of reservoir temperature during matrix acidizing treatments [23]. The lines represent the injection rate required to achieve wormhole formation for various acid systems including aqueous HCl, organic acids HFc and HAc, and emulsified HCl. The shaded box represents typical injection rates used in matrix stimulation treatments. The results show that aqueous HCl is unable to achieve effective wormhole formation without requiring excessive injection rates (above the typical ranges that would avoid fracturing the formation). Under the same conditions, emulsified HCl can effectively stimulate the formation even at higher temperatures where aqueous systems would not be effective. The results demonstrate that emulsified acids generate more effective acid dissolution channels under typical treatment conditions, which enables increased hydrocarbon production.

While emulsified acid systems are more effective than aqueous HCl for carbonate acidizing applications, they often require treatments to be pumped at lower than optimal injection rates due to relatively high friction pressures. The emulsified acid viscosities typically vary between 2.5 and 4.5 cp at 800 s^{-1} and 121–176°C, which is an order of magnitude higher than the 0.25 and 0.5 cp viscosity of aqueous HCl over the same temperature range [25]. Viscosities around 3 cp were reported for acid microemulsion systems [9]. Other investigators have reported significantly lower viscosities of about 0.1 cp at 400 s^{-1} for microemulsions [38]. The authors are not aware of friction loss measurements for the flow of acid microemulsion systems under typical oilfield conditions. Therefore, friction studies could be an interesting area for further investigation to assess the potential of acid microemulsions to mitigate this operational limitation and provide a technical advantage over macroemulsion systems commonly used in carbonate acidizing applications.

7. Summary

Microemulsion technologies have unique properties that make them applicable in the oil and gas industry. Of particular value for carbonate acidizing is their ability to reduce the mobility of the dispersed phase, in this case hydrochloric acid, without a proportionate increase in the fluid viscosity. Microemulsions provide effective acid diffusion coefficients that are two orders of magnitude lower than aqueous HCl. This reduction in the diffusion coefficient is important in carbonate acidizing because the rate of dissolution is mass transfer limited, meaning it can be controlled by changing the rate of convection and/or the rate of acid diffusion.

This research evaluated acid microemulsions with respect to the phase behavior, effective diffusivity, and corresponding impact on the dissolution phenomenon occurring during carbonate acidizing. Results demonstrate that acid microemulsions are effective at creating highly-conductive dissolution channels, or wormholes, in carbonate porous media. Furthermore, the dissolution structure and corresponding volume of fluid required for a given depth of stimulation are governed by the Damköhler number for transport and reaction. Due to the significant reduction in the effective diffusivity of the dispersed acid phase and the dependence of the dissolution phenomenon on the Damköhler number, acid microemulsions provide improved performance compared to aqueous acids, particularly at low injection rates and high reservoir temperatures.

Although research demonstrates that acid microemulsions are effective carbonate acidizing fluid systems, the authors are not aware of microemulsions prepared with acid as the dispersed phase being used for treatments in the field. Emulsified acids ("macro"emulsions) are commonly used for carbonate acidizing and provide a similar reduction in the effective diffusion coefficient as observed for acid microemulsions. These macroemulsion systems often constrain operating conditions due to high friction pressures. Friction studies could be an interesting area for future acid microemulsion research.

Author details

Christopher N. Fredd[1*], Mark L. Hoefner[2] and H. Scott Fogler[3]

*Address all correspondence to: cfredd@slb.com

1 Schlumberger, Dubai, UAE

2 ExxonMobil, Houston, TX, USA

3 University of Michigan, Ann Arbor, MI, USA

References

[1] Penny GS, Pursley JT, Holcomb D. The Application of Microemulsion Additives in Drilling and Stimulation Results in Enhanced Gas Production. In: 2005 SPE Production and Operations Symposium; 17–19 April 2005; Oklahoma City, OK. DOI: 10.2118/94274-MS

[2] Agee DM, Wirajati AY, Schafer L, Grant G, Garnier A, Thouvenin E, Wijanarko A. Post-Fracturing Fluid Recovery Enhancement with Microemulsion. In: 2010 SPE International Symposium and Exhibition on Formation Damage Control; 10–12 February 2010; Lafayette, Louisiana. DOI: 10.2118/128098-MS

[3] Zelenev AS, Zhou H, Ellena L, Penny GS. Microemulsion-Assisted Fluid Recovery and Improved Permeability to Gas in Shale Formations. In: 2010 SPE International Symposium and Exhibition on Formation Damage Control; 10–12 February 2010; Lafayette, Louisiana. DOI: 10.2118/127922-M

[4] Nasr-El-Din HA, Shedd, D, Germack D, Penny G. Evaluation of Methods to Improve Matrix Stimulation of Horizontal Carbonate Wellbores. In: Abu Dhabi International Petroleum Exhibition and Conference; 9–12 November 2015; Abu Dhabi, U.A.E. DOI: 10.2118/177934-MS

[5] Brege J, El Sherbeny WIA, Quintero L, Jones TA. Using Microemulsion Technology to Remove Oil-based Mud in Wellbore Displacement and Remediation Applications. In: The North Africa Technical Conference and Exhibition; 20–21 February 2012; Cairo, Egypt. DOI: 10.2118/150237-MS

[6] Suniga PT, Fortenberry R, Delshad M. Observations of Microemulsion Viscosity for Surfactant EOR Processes. In: SPE Improved Oil Recovery Conference; 11–13 April 2016; Tulsa, OK. DOI: 10.2118/179669-MS

[7] Humphry KJ, van der Lee M, Southwick JG, Ineke EM, van Batenburg DW. Microemulsion Flow in Porous Media: Implications for Alkaline-Surfactant-Polymer Flooding. In: SPE Enhanced Oil Recovery Conference; 2–4 July 2013; Kuala Lumpur, Malaysia. DOI: 10.2118/165233-MS

[8] Hartley R, Bosma MGR. Fracturing in Chalk Completions. Mint: Journal of Petroleum Technology. Jan 1985; 73–79.

[9] Hoefner ML, Fogler HS. Role of Acid Diffusion in Matrix Acidizing of Carbonates. Mint: Journal of Petroleum Technology. Feb 1987; 203–208.

[10] Hoefner ML, Fogler HS. Pore Evolution and Channel Formation During Flow and Reaction in Porous Media. Mint: AIChE Journal. 1988; 34(1): 45–54.

[11] Fredd CN, Fogler HS. Optimum Conditions for Wormhole Formation in Carbonate Porous Media:Influence of Transport and Reaction. Mint: SPE Journal. 1999; 4(3): 196–205.

[12] McCload HO. Matrix Acidizing. Mint: Journal of Petroleum Technology. Dec 1984; 2055–2069.

[13] Nierode DE, Williams BB. Characteristics of Acid Reaction in Limestone Formations. Mint: SPE Journal. Dec 1971; 406–418.

[14] Daccord G, Touboul E, Lenormand R. Carbonate Acidizing: Toward a Quantitative Model of the Wormholing Phenomena. Mint: SPE Production Engineering. Feb 1989; 63–68.

[15] Mostofizadeh B, Economides MJ. Optimum Injection Rate From Radial Acidizing Experiments. In: paper SPE 28547 presented at the SPE 69th Annual Technical Conference and Exhibition; 25–28 September 1994; New Orleans, LA.

[16] Buijse MA. Understanding Wormholing Mechanisms Can Improve Acid Treatments in Carbonate Formations. In: paper SPE 38166 presented at the European Formation Damage Conference; 2–3 June 1997; The Hague, The Netherlands.

[17] Wang Y, Hill AD, Schechter RS. The Optimum Injection Rate for Matrix Acidizing of Carbonate Formations. In: paper SPE 26578 presented at the Annual Technical Conference and Exhibition; 3–6 October 1993; Houston, TX.

[18] Lund K, Fogler HS, McCune CC, Ault JW. Acidization – II. The Dissolution of Calcite in Hydrochloric Acid. Mint: Chemical Engineering Science. 1975; 30: 825.

[19] Lund K, Fogler HS, McCune CC. Acidization – I. The Dissolution of Dolomite in Hydrochloric Acid. Mint: Chemical Engineering Science. 1973; 28: 691–700.

[20] Gardner TR, Dill WR, Ford WGF, King KL. Well Acidizing Compositions and Method. Patent US 5034140 A. November 1989, Lapsed.

[21] Chang F, Fogler HS. Chapter 4:Carbonate Acidizing. In: Acid Stimulation, SPE Monograph Series. Vol. 26, 2016, p. 85-115.

[22] Fredd CN, Fogler HS. Influence of Transport and Reaction on Wormhole Formation in Porous Media. Mint: AIChE Journal. 1998; 44(9): 1933–1949.

[23] Fredd CN. Dynamic Model of Wormhole Formation Demonstrates Conditions for Effective Skin Reduction During Carbonate Matrix Acidizing. In: paper SPE 59537 presented at the 2000 SPE Permian Basin Oil and Gas Recovery Conference; 21–23 March 2000; Midland, TX.

[24] Fredd CN. Fracture Stimulation Process for Carbonate Reservoirs. Patent US 6749022 B1. June 2004.

[25] Navarrete RC, Holms BA, McConnell SB, Linton DE. Emulsified Acid Enhances Well Production in High Temperature Carbonate Formations. In: paper SPE 50612 presented the 1998 SPE European Petroleum Conference; 20–22 October 1998; The Hague, The Netherlands. DOI: 10.2118/50612-MS

[26] Melendez MG, Pournik M, Zhu D, Hill AD. The Effects of Acid Contact Time and the Resulting Weakening of the Rock Surfaces on Acid Fracture Conductivity. In: paper SPE 107772 presented at the 2007 SPE European Formation Damage Meeting; 30–1 May–June 2007; Scheveningen, The Netherlands. DOI: 10.2118/107772-MS

[27] Levich VG. Physicochemical Hydrodynamics. Prentice-Hall; 1962; Englewood Cliffs, NJ.

[28] Boomer DR, McCune CC, Fogler HS. Rotating Disk Apparatus for Reaction Rate Studies in Corrosive Liquid Environments. Mint: Review Scientific Literature. 1972; 43(2): 225–229.

[29] McCune CC, Fogler HS, Kline WE. An Experimental Technique for Obtaining Permeability-Porosity Relationships in Acidized Porous Media. Mint: Industrial Engineering Chemistry Fundamental. 1979; 18(2): 188.

[30] Jasti JK, Fogler HS. Application of Neutron Radiography to Image Flow Phenomena in Porous Media. Mint: AIChE Journal. 1992; 38(4): 481–488.

[31] Fredd CN, Lindsay JT, Fogler HS. Neutron Transmission Tomography Applied to Reactive Dissolution Through Percolating Porous Media. In: Chaouki J, Larachi F, Dudukovic MP, editors. Non-Invasive Monitoring of Multiphase Flows. Elsevier Science Publishing Co., 1997, p. 185. New York, NY.

[32] Hoefner ML, Fogler HS. Effective Matrix Acidizing in Carbonates Using Microemulsions. Mint: Chemical Engineering Progress. May 1985; 40–44.

[33] Hoefner ML. Matrix Acidizing in Carbonates Using Microemulsions. In: Ph.D. Thesis, The University of Michigan Department of Chemical Engineering: 1987. Ann Arbor, MI.

[34] de Rozieres J, Chang FF, Sullivan RB. Measuring Diffusion Coefficients in Acid Fracturing Fluids and Their Application to Gelled and Emulsified Acids. In: paper SPE 28552 presented at the SPE Annual Technical Conference and Exhibition; 25–28 September 1994; New Orleans, LA.

[35] Nasr-El-Din HA, Solares JR, Al-Mutairi SH, Mahoney MD. Field Application of Emulsified Acid-Based System to Stimulate Deep, Sour Gas Reservoirs in Saudi Arabia. In: paper SPE71693 presented at the 2001 SPE Annual Conference and Exhibition; 30–03 September–October 2001. DOI: 10.2118/71693-MS

[36] Rotondi M, Pace H, Sobernheim DW, Malonga H, Batmaz T, Pounga F, Obondoko G, Amare P. Application of Different Stimulation Techniques, Multistage Proppant and Acid Fracturing Operations Offshore Congo. In: paper SPE 134905 presented at the SPE Annual Technical Conference and Exhibition; 19–22 September 2010; Florence, Italy. DOI: 10.2118/134905-MS

[37] AL-Hassan A, Anthony E, El-Aziz SA, Abdel-Basset M, Mokhtar A, Sheikh B. First Acid Fracturing Unlocks the Production Potential of Tight Tuba Carbonate Reservoir, North Kuwait. In: paper SPE 175230 presented at the SPE Kuwait Oil & Gas Show and Conference; 11–14 October 2015; Mishref, Kuwait. DOI: 10.2118/175230-MS

[38] Pandava Aum PT, Souza TN, GurgelAum YKP, Dantas TNC, DantasNeto AA. New Acid O/W Microemulsion Systems for Application in Carbonate Acidizing. Mint: International Journal of Advanced Scientific and Technical Research. 2016; 6(1): 182–196.

Chapter 4

Microemulsion Route for the Synthesis of Nano-Structured Catalytic Materials

Tajamal Hussain and Rabia Batool

Additional information is available at the end of the chapter

http://dx.doi.org/10.5772/66183

Abstract

Owing to their unique properties, use of microemulsion-based synthetic techniques for the generation of shape-controlled nanocatalyst is an area of great current interest. Nanocatalysts of any specific shape, morphology, surface area, size, geometry, homogeneity and composition are widely being prepared using the soft techniques of microemulsion. Easy handling, inexpensive equipment and mild reaction conditions make microemulsion an attractive reaction medium. Herein, a nanosized precursor reactant can be incorporated, leading to the formulation of a highly monodispersed metal nanoagglomerate with controlled size, shape and composition. Several factors such as presence of electrolyte, molar ratio of water to surfactant, nature and concentration of surfactant and solvent, size of water droplets and concentration of reducing agents influence the size of the nanoparticles. The reverse micelle method can be used for the fabrication of several nanosized catalysts with a diverse variety of suitable materials including silica, alumina, metals (e.g. Au, Pd, Rh, Pt), metal oxides, etc. The morphology, size distribution and shape of the nanocatalysts make them useable for a wide range of applications, for example, fuel cells, electrocatalysis, photocatalysis, environmental protection, etc. The recovery of nanoparticles from the reaction mixture is a challenge for the researchers. This chapter discusses the preparation of nanoparticles using microemulsion techniques, widely being used for the synthesis of nanocatalysts from a wide range of materials.

Keywords: microemulsion, nanoparticles, catalyst, synthesis, nanoagglomerate

1. Introduction

A single-phase system formed by the addition of a surfactant to a mixture of two immiscible liquids in large amounts, which is homogeneous macroscopically but heterogeneous microscopically, have been fascinating the world of science and technology due to their unique properties such as ability to solubilize both oil- and water-soluble compounds, low viscosity, decreased interfacial tension, optical clarity, etc. [1–3]. Unlike, normal emulsions, they exhibit a high thermodynamic stability as proposed by Ruckenstein [4]. These isotropic systems, consisting of oil droplets dispersed in water (O/W) or water droplets dispersed in oil medium (W/O) or a bicontinuous phase of these two components were termed as 'microemulsions' in 1959 by Schulman [5, 6].

A microemulsion basically consists of a hydrocarbon, water and a surfactant (a molecule having a hydrophobic tail and hydrophilic head ends) [7, 8]. After a specific concentration of surfactant (critical micelle concentration), aggregation of molecules takes place and micelles are formed. In case of W/O microemulsion, where formation of micelles takes place in an organic medium, whose hydrophobic (nonpolar) tail is outside the core and interact with the hydrocarbons, whereas its hydrophilic (polar) head is in the core; the aggregates are referred to as 'reversed micelles'. The core of W/O microemulsion, an active candidate as a nanoreactor in which several chemical reactions can be carried out, is of great interest [9]. Sometimes, a cosurfactant, mostly a medium-chain alcohol, is also used in addition to the surfactant, which enhances the system's stability and entropy by increasing the fluidity and decreasing the interfacial tension between water and oil molecules, thus easing the microemulsion formation [10]. The stability of a microemulsion can also be enhanced by changing the reaction parameters such as pressure, temperature or concentration of reagents or by the addition of a salt [11].

Depending on the ratio of individual constituents of a microemulsion, it is highly sensitive to a slight change in temperature [12]. Microemulsions are a subject of extensive research due to their wide range of applications in several fields, including cosmetics [13], lubricants [14], as pesticides in agrochemicals [14, 15], textile industry as finishing and dyeing materials, electrocatalytic, organic and photochemical reactions [16, 17], liquid membranes, fuels, detoxification of environment [18, 19], pharmaceutical industry [15, 20, 21], corrosion inhibitors [14], detergents [2], as oil flavors in the food industry [2, 22], bio-separation [23], oil recovery [24], etc. Nanoparticles having unique properties, finding their potential use in homogeneous and heterogeneous electrocatalysis, fluorescence probes, thermoelectric transport, bioassays, optical nanosensors, biotechnology, bio-imaging, aerosols, etc. are also extensively being synthesized using microemulsion technique. This chapter presents an overview of different microemulsion techniques used for the synthesis of nanocatalytic materials, factors affecting the reaction rate as well as the nanoparticle size and their applications in various fields.

2. Synthesis of nano-catalysts

In 1972, Gault and coworkers employed microemulsions, a novel technique for the synthesis of catalytic materials for the very first time. Owing to its thermal stability, specific composition and structure, very small droplet size and transparency, microemulsion, being a suitable medium for the preparation of metal-based nanocatalysts are also reported for the formation of bimetallic nanoparticles of any desired composition and controlled particle size. Metal nanoparticles stabilized in a microemulsion medium were first reported by Boutonnet [16].

Basically, the catalytic ability of a metal-based nanocatalyst depends on the size of precursor metal particles; smaller the particle size, higher will be the activity, as the surface area of the metal particles is increased due to increasing number of atoms on its surface. Microemulsion is a promising method leading to the synthesis of catalysts at nanoscale with a narrow size distribution. Varying the size of reversed micelles leads to control over the size of catalytic particles, ranging from 1–100 nm. The composition of nanocatalysts as prepared by this technique corresponds to the initial concentration of metal precursor used for their synthesis. In contrast with the conventional methods, which require a high temperature for the synthesis of nanoparticles, reverse micelle method can be carried out at room temperature with a more précised size of nanocatalysts. Water in oil microemulsion is of higher interest in this regard. In W/O microemulsion, water droplets are dispersed in oil. The surfactant molecules surround the water particles with their polar ends facing the water particles and nonpolar ends towards the oil phase. Metal precursors are present as electrolytes in the form of metal salts in the water core of the aggregations. Either this system can be mixed with another microemulsion containing precipitating or reducing agents (hydrogen, sodium borohydride, hydrazine, etc.), or a precipitating agent can be separately added to this microemulsion system [7]. This scheme is represented in **Figure 1**. A homogeneous dispersion of precipitating agent is achieved in the former method. Nucleation reaction takes place by the collision of these agents with the precursor metal particles. Nuclei formed in this process grow to form the final form of nanocatalyst. Since oil is a continuous phase, yield of resulting catalyst can be increased by increasing the amount of oil in the system [25]. Eastoe and coworkers [25] introduced the use of environment-friendly supercritical carbon dioxide instead of oil as a continuous phase.

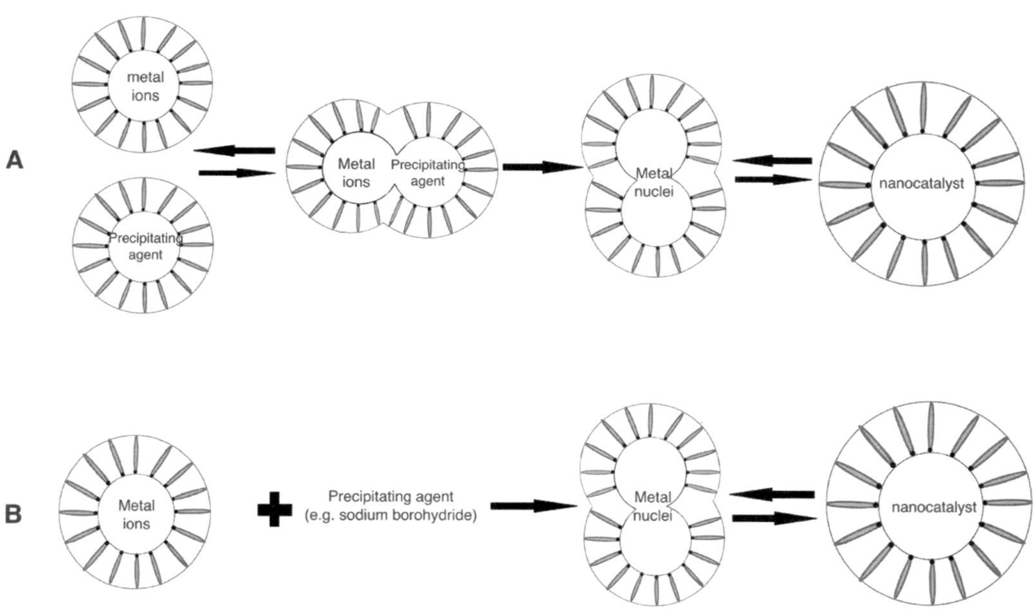

Figure 1. Nanocatalyst synthesis in microemulsion medium: (A) Mixing of two microemulsions; (B) addition of precipitating to microemulsion.

The nanocatalysts formed through microemulsion route are then passed through after-treatments such as reduction, washing, stabilization, calcination, etc. depending on the fields in which they are to be used [26]. The main hindrance, which prevents the approach of reactants to the surface of catalyst, is the presence of surfactant molecules. So, it must be removed from the reaction medium. Removal of nonionic surfactant molecules from the medium is easier compared to ionic surfactant. Thus, for the catalysts which are destined to be used at low as well as high-temperature liquid phase catalysis, surfactant should be removed; this can be achieved through calcination and washing. However, in some cases, presence of surfactant molecules on the surface of metal-based nanocatalyst particles may improve its selectivity. Moreover, if microemulsion-prepared metal nanocatalysts come in contact with air, they may get oxidized. To avoid this problem, reduction of nanocatalyst in hydrogen must be carried out before using it for a catalytic process.

Microemulsion has also been explored for synthesis of bimetallic nanocatalysts (catalysts containing two different metal particles). The addition of a second metal particle in the system improves its catalytic activity. Two different pathways can be employed for this purpose; either separate aqueous solutions of both precursors or an aqueous solution having both the precursor metals can be added to the microemulsion medium. However, the selection of route does not affect the particle size of product [27].

To increase the efficiency of a nanocatalyst, encapsulation of metal particles should be avoided to maximize the surface area of catalyst by creating them on a support [28]. A combination of microemulsion and sol-gel method can be employed for the synthesis of metal oxide catalysts coated with metal particles. Illustration of this method is given in **Figure 2**. A sol-gel technique is used to prepare a porous metal oxide via reactions between metal alkoxide and water. In microemulsion, it prevents sintering and agglomeration of precursor metal particles in the medium.

Microemulsion coupled with deposition precipitation method is also being used to support active catalysts over surface of nanomaterial. In deposition precipitation method, particles of metal deposited on the surface act as active catalysts. The surface of nanomaterials acts as a support for the metal catalyst. This will help in improving the catalytic ability as well as ther-

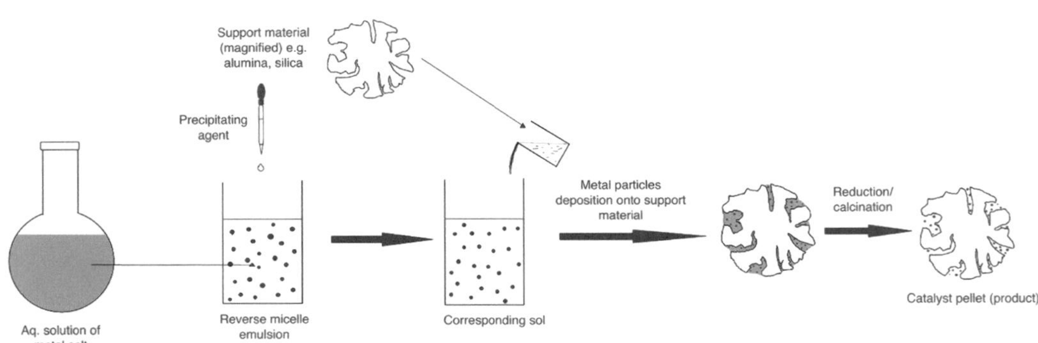

Figure 2. Microemulsion route for the synthesis of supported metal catalysts.

mal stability of the catalyst [29]. Such a three-way catalyst (TWC) has been prepared by Shah et al. [30], who supported CuO particles on $CesZrO_2$. The doping of CuO activated the highly stable catalyst at lower temperature.

3. Factors affecting the particle size

3.1. Concentration of water

Water content in the microemulsion goes hand in hand with particle size of the catalyst. Increasing the water content increases the water-to-surfactant ratio, increasing the particle size of metal catalysts (see **Table 1**). Micellar size is decreased at lower water contents, which results in homogeneous dispersion of relatively smaller particle size. At low amount of water, the tightly packed water molecule surrounded by surfactant molecules results in the production of more stabilized and monodispersed as well as smaller catalyst particles. This means that for the synthesis of catalysts having a smaller size, amount of water in the microemulsion should be reduced [36–38]. This effect was studied by Wongwailikhit and Horwongsakul. They selected reverse micellar route for the synthesis of monodispersed, spherical iron (III) oxide nanocatalysts. Particle size of the nanocatalyst was found to be strongly dependent on molar ratio of water to surfactant. Decrease in concentration of water in the system favored the decrease in particle size, thus enhancing the catalytic activity of product [39].

3.2. Intermicellar exchange

Microemulsion synthesis of nanocatalysts depends strongly on the intermicellar exchange. If the rate of exchange of materials between reverse micelles is slow, a few catalyst particles with relatively larger particle size are obtained. In contrast, higher exchange rate leads to the synthesis of a large number of nanoparticles having smaller particle size [40]. As

Serial #	Nano-catalyst	Water to surfactant ratio (w)	Size of nanocatalyst (nm)	Reference
1	Cu_2S	2	8.09	[31]
		3	69.2	
2	TiO_2	3	5.3	[32]
		5	6.9	
3	$MnCO_3$	6.25	71	[33]
		7.5	89	
4	$MgAl_2O_4$	2	13	[34]
		8	17	
5	CdS	5	6.2	[35]
		10	10.80	

Table 1. Effect of water-to-surfactant ratio on size of nanocatalyst.

reported by Bagwe et al. [41] who synthesized silver chloride nanoparticles in microemulsions, the intermicellar exchange further depends on the water content, choice of continuous phase and precursor metal particles in the nanodroplets. At greater water-to-surfactant molar ratio, water molecules exists in free state in the reaction medium; this increase in medium's fluidity enhances the intermicellar exchange rate and forms smaller nanocatalyst size owing to their rapid nucleation. The intermicellar exchange is also weakly affected by the size of cations. It is well known that cationic size decreases with increasing valency, reducing the distance between surfactant heads and forming a tightly packed layer of surfactant. Hence, larger cations affect the formation of nanocatalysts by reducing the intermicellar exchange. Another factor affecting materials exchange is the continuous phase (solvent) present in the microemulsion medium. The penetration of continuous (oil) phase into the surfactant tail becomes difficult as the chain length of the oil is increased; however, oil molecules align themselves parallel to the tail, decreasing the interaction between continuous phase, that is, solvent and surfactant. This deficiency of solvent molecules allows the surfactant tails to interact with each other. These two effects combine and result in an increased intermicellar exchange.

3.3. Reducing or precipitating agent

The nature and concentration of precipitating or reducing agent is a strong function of the microemulsion synthesized metal nanocatalyst size. As reported by Boutonnet et al., increase in the content of hydrazine, N_2H_4, which is an efficient reducing agent, increases the particle size of metal nanocatalysts [42]. In this work, it is reported that N_2H_4 carries out reduction more effectively and forms comparatively smaller nanoparticles as compared to when pure hydrogen is used for reduction.

A research carried out by Solanki [43] showed that nature of reducing agent in microemulsion technique affects the particle size of nanocatalysts. They used two different reductants, that is, hydrazine and sodium borohydride for the synthesis of silver (Ag) nanocatalysts keeping the water-to-surfactant ratio constant. Sodium borohydride, being a stronger reducing agent, catalyzes the nucleation and growth of particles, resulting in greater particle size. Therefore, hydrazine is more preferable for the synthesis of nanocatalysts with reduced particle size.

3.4. Nature of surfactant

Particle size of a nanocatalyst is also affected by changing the concentration of surfactant in the emulsion keeping the water and oil content constant. Ionic and van der Waals forces hold the aggregates formed in the presence of surfactant. By reducing the amount of surfactant, number of droplets also decreases. This causes an increase in ions per droplet and hence the particle size also increases. However, the catalyst particles are formed in the nuclei, not in the droplet. It is known that the dynamic microemulsion system synthesizes nanocatalyst in two steps: nucleation and aggregation. Presence of surfactant keeps the particle growth neither too slow nor too fast. Particle size is determined by the number of surfactant molecules surrounding it [44, 45].

Surfactants being used in microemulsion-mediated synthesis of nanocatalysts include different nonionic, anionic and cationic surfactants. Anionic surfactant AOT (sodium bis(2-ethylhexyl) sulfosuccinate), employed for the synthesis of monodispersed and highly stable metallic (Au, Ni, Cu) nanoparticles as well as metal oxides and sulfides catalysts with very fine particle size strongly bonded through electrostatic interactions, is the most important surfactant in this regard [46].

3.5. Addition of a cosurfactant

Providing same particle size, addition of a cosurfactant in the microemulsion reaction medium during the synthesis of a bimetallic nanocatalysts assists the complete reduction of precursor metal ions to form well-alloyed monodispersed nanoparticles with desired phase composition. The presence of a cosurfactant incorporates more amounts of metal particles in the catalyst, owing to the increased rigidity of micellar interface. The cosurfactant molecules solubilize in the tail region of surfactant, the heads of surfactant come close to each other, making no effect on the water droplet size, which determines the particle size of nanocatalysts [47]. Increasing the chain length of cosurfactant stabilizes the nanocatalyst particles formed, leading to a decreased average particle size. This effect was shown by He [48] and coworkers, studying the size of nanocatalysts with changing cosurfactant in the microemulsion medium (see **Table 2**).

3.6. Addition of electrolytes

Several electrolytic species such as metal salts, which have a great impact on aggregation of micelles, the solubility and dissociative degree of the surfactant, may also be added to the microemulsion medium used for the synthesis of nanocatalysts [49]. For instance, shape of catalyst particles in water-AOT-isoctane microemulsion was altered by the addition of sodium chloride salt in the system, where cubical, cylindrical or trigonal catalytic particles were obtained [50].

3.7. Reactant concentration

The concentration of reactants in the microemulsion medium strongly affects the amount, size and shape of the nanocatalysts thus synthesized. The particle size of silver nanoparticles as prepared by Lisiecki et al. [51] was observed to decrease with decreasing concentration of reducing agent (hydrazine). An increase in reactants concentration (PbS, $NaBH_4$) increased the amount of final product, catalysts particles [52, 53].

Serial #	Co-surfactant	Size of nanocatalyst (nm)
1	1-Heptyl alcohol	47.1
2	1-Amy alcohol	51.5
3	1-Hexyl alcohol	53.8
4	1-Butyl alcohol	63.4

Table 2. Effect of surfactant on size of nanocatalyst.

An important edge a microemulsion provides is its potential to control the shape (onion-like, core shell, etc.) of the synthesized nanocatalysts by changing the initial concentration of reactants added to the system [54, 55]. Core with uniform thickness of particles can be obtained by using rigid films or introducing low concentration of reactants in the reaction mixture providing one of the reactants is in excess amount. Using stoichiometric and relatively higher amounts of reactants and addition of a second reactant to the reaction mixture produce a core-shell structure [55, 56]. Efforts are also being done for the polymer encapsulation of microemulsion synthesized nanocatalyst particles either by in situ polymerization by making use of polymerized surfactants [57] or by enfolding after the extraction of nanocatalysts [58].

4. Applications of microemulsion synthesized nanocatalysts

The nanocatalysts prepared by using microemulsion systems have been reported to be used in many catalytic processes including reforming, oxidation/hydrogenation of hydrocarbons, reduction of nitrogen oxide, fuel cells, Fischer-Tropsch method, production of hydrogen, oxidation of carbon monoxide, breaking down of organic species, etc. [26] (see **Table 3**). Details are given below.

4.1. Photocatalysis

Microemulsion technique produces better nanostructured photocatalyts. Photocatalysis is a process used for the breakdown of organic matter present in wastewaters. Titanium dioxide, being a semiconductor, is widely being employed in this process. With the absorption of photon whose energy is equal to semiconductor's band gap, the catalysis begins. Titania either accepts or donates electrons to the species present in the surrounding medium. Yan and coworkers [58] synthesized one such catalyst via microemulsion route, which enhanced its catalytic activity. Fuerte et al. [65] studied the effect of microemulsion route on photocatalytic activity of mixed oxides nanoparticles containing titanium (Ti) and tungsten (W) for degradation of toluene. In the microemulsion medium, tungsten doping leads to the formation of charge-trapping centers, leading to enhanced catalytic activity of the nanocatalyst. Another core/shell catalytic system containing ZnS and CdS are known to be applicable for the photolytic production of hydrogen from water [66].

Photocatalytic decomposition of 4-nitrophenol, a toxic compound used in pesticide production, is carried out over microemulsion-synthesized titania nanocatalysts [67]. Organic contaminants present in water can be degraded under UV/visible light using an active single-phase $LaNiO_3$ as prepared by a single reversed microemulsion process [68].

4.2. Storage of hydrogen

Microemulsion-synthesized nanosized bimetallic catalysts find their applications to enhance the reaction rate of formation of carbon nanofibers, which are used for storage of hydrogen. The nanofibers as produced by this technique show a uniform distribution of very small

Serial#	Nanocatalyst	Application	Reference
1	Pt	Oxygen reduction reaction	[49]
2	Pt-Ru	Oxidation of methanol	[50]
3	TiO_2	Photodegradation of methyl orange	[51]
4	Pd/CNT	Hydrogenation of olefins	[52]
5	Pd-Au/carbon	Electrooxidation of sodium borohydride	[53]
6	Ag/Al_2O_3	Reduction of nitro aromatic compounds	[37]
7	Barium hexaaluminate	Combustion of methane	[54]
8	Fe/Cu/La	Fischer-Tropsch catalyst	[55]
9	MoO_3-K_2O/CNT	Synthesis of higher alcohols	[56]
10	Mn-containing mixed oxides	Catalytic oxidation of toluene	[57]
11	Rh/SiO_2	CO hydrogenation	[58]
12	Pt-in-CeO_2	Water-gas shift reaction	[59]
13	Pd-Cu-O/SiO_2	Oxidative carbonylation of phenol	[60]
14	Pt-Ru-Sn/BDD	Electro-oxidation of alcohol	[61]
15	Pd-Co-Au	Oxygen reduction	[62]
16	Pt/carbon	Hydrogenation of nitro compounds	[63]
17	Ni-Fe/Al_2O_3	Hydrogen rich gas production	[64]

Table 3. Applications of microemulsion-synthesized nanocatalysts.

particle size. A Fe-Cu nanocatalyst was synthesized by Marella et al. [69] to study the formation of nanofibers through the decomposition of ethylene.

4.3. Aquathermolysis reaction

The limited use of heavy oils due to their high viscosity can be improved by reducing its viscosity. For this purpose, aquathermolysis, physiochemical reactions of oil with water or steam at temperature range between 200 to 300°C is an important reaction [70]. However, a major problem observed in this reaction was the formation of large molecules by Van der Waals or hydrogen-bonded hetero-atoms (O, N, S) present in the heavy oil [71]. To overcome this difficulty, a catalyst can be used which catalyzes the breakdown of large oil particles into smaller ones, increasing the amount of saturates and decreasing the concentration of resins and asphaltenes in the oil [72]. The enhanced catalytic activity of microemulsion-synthesized nanocatalysts due to their smaller particle size makes them an interesting candidate to be used in this viscosity reduction method [73].

4.4. Reduction of nitrogen oxides

In the presence of excess oxygen, mixed metal nanocatalysts such as perovskites [74] are generally used for the reduction of nitrogen oxides (Eq. (1)). Microemulsion synthesis enhances the activity of nanocatalysts by decreasing their particle size and increasing their surface area. Homogeneously mixed precursor metal particles in the microemulsion medium lead to the formation of nanocatalysts at elevated temperatures:

$$2NO + O_2 \xrightarrow{CaTiO_3} 2\,NO_2 \tag{1}$$

Reduction of nitric oxide by carbon monoxide using Lanthanum-based nanocatalysts [75] synthesized in microemulsion medium is also reported. The activity was compared with the conventionally prepared nanocatalysts. The enhanced surface area of microemulsion-synthesized nanocatalysts increased the rate at which oxidation reaction occurred.

In gas turbines, nitrogen oxide formation should be minimized by reducing the temperature at which catalytic combustion is carried out. For this purpose, hexaaluminate can be used as a catalyst. This catalyst having a higher surface area, when prepared via microemulsion technique, is an active candidate to be used at low temperatures [76].

4.5. Water-gas shift reaction

In water-gas shift (WGS) reactions, carbon dioxide and hydrogen are produced by reacting carbon monoxide with water. These reactions take place in industries to increase the amount of hydrogen in syngas (a fuel gas produced as a result of partial oxidation of hydrocarbons). These reactions can be catalyzed using ceria (CeO_2) encapsulated Pt nanoparticles synthesized in microemulsion medium (Eq. (2)) [77]:

$$CO + H_2O \xrightarrow{CeO_2} H_2 + CO_2 \tag{2}$$

4.6. Fischer-Tropsch synthesis

Use of nanocatalysts based on metal prepared through microemulsion technique in Fischer-Tropsch (FT) synthesis (conversion of hydrogen and carbon monoxide mixture into liquid hydrocarbon, Eq. (3)) is of great interest. Due to their low cost, selectivity and activity towards FT synthesis, low deactivation rate, supported cobalt catalysts are highly recommended to be used in FT synthesis [78]. Re supported Co catalysts were synthesized and deposited on titania. Using microemulsion method, smaller particle size of these catalysts was obtained whose catalytic activity in FT reactions was very much higher as compared to those prepared by other methods. Carbon nanotubes (CNT), when used as a catalyst carrier-support for Co nanoparticles, provides a higher reproducibility and higher degree of dispersion of catalyst [79]. Moreover, the catalytic activity of Co in FTS depends on its size distribution [80]:

$$(2n+1)H_2 + nCO \xrightarrow{Co} C_nH_{2n+2} + n\,H_2O \tag{3}$$

http://dx.doi.org/10.5772/66183

4.7. Electrocatalysis

Metal particles dispersed on carbon electrode are most widely used catalysts for electrolytic reactions. Xiong and He [81] reported the use of tungsten oxide and tungsten nanoparticles in electrocatalysis. They reported that surfactant must be removed from the surface of catalysts to make them useable at elevated temperatures. Iron-based nanocatalysts prepared in microemulsion medium found application in the electrocatalytic reduction of hydrogen peroxide (Eq. (4)) [82]. Cathodic catalysts to be used in fuel cells are also reported to be synthesized via microemulsion route [83]. Yet, the microemulsion-synthesized nanocatalysts exhibit unsatisfactory long-term stability [84]:

$$H_2O_2 + 2H^+ \overset{Fe}{\Leftrightarrow} 2H_2O \tag{4}$$

4.8. Combustion/oxidation/hydrogenation of hydrocarbons

Hexaaluminate is a catalyst used widely for catalytic combustion of methane. To enhance the activity of a catalyst, it is highly recommended to increase its surface area. Microemulsion route is proven to be a confirmed way to increase surface area of catalysts by decreasing their particle size [85]. The homogeneity of particles in microemulsion medium is increased, leading to the formation of catalyst at elevated temperatures, hence increasing its surface area. Hydrothermal treatment of microemulsion during synthesis of catalyst affects the effective surface area of catalyst. Particles having a smaller size over the range of approximately 2 nm ignite the process of reaction [86]. Silica- and alumina-coated metal particles such as platinum and nickel particles are also known to be used for complete and partial catalytic oxidation of methane (Eq. (5)); the high selectivity of this reaction lies in diffusion of methane particles to the catalyst surface through the silica coating [87–89]. The coating prevents the detachment of precursor metal particles from the support material, growth of carbon fibers and conversion of metal particles into their oxides and makes the nanocatalysts more stable and reactive. Synthesizing these catalysts in microemulsion route minimizes the size of particles rendering their activity high:

$$CH_4 + 2O_2 \xrightarrow{Pt/Ni} CO_2 + 2H_2O \tag{5}$$

Trimetallic nanocatalysts [90] prepared in microemulsion are more effective as compared to bimetallic catalysts for the oxidation of methanol as well as ethanol. Microemulsion synthesis overcomes the problems faced in the oxidation process, that is, poisoning by carbon monoxide, slow process of oxygen reduction [91, 88]. Carbon nanotube supported Pd, Rh and bimetallic (Pd/Rh) nanocatalysts are also employed as very effective catalysts for hydrogenation of olefins (Eq. (6)) [59]:

$$RCH = CH_2 + H_2 \xrightarrow{Pd/CNT} RCH_2CH_3 \tag{6}$$

A microemulsion based on a less toxic solvent, super critical CO_2, is employed by Meric and coworkers [89] for synthesis of palladium nanocatalysts. Due to capacity of this solvent to dissolve hydrogen and oxygen gases, it is highly recommended to be used in liquid-phase reactions as a medium. It also enhances the rate at which the reaction proceeds. The effect of Pd nanocatalysts on citral (3,7-dimethyl-2,6-octadienal) hydrogenation was studied by Meric. Using nanocatalysts dispersed and stabilized in microemulsion, approximately 68% unsaturated aldehyde (citeranol) was obtained. However, saturated aldehyde was gained as a main product in liquid or vapor phase reactions. This difference lies in the presence of palladium enclosed by micelle particles. Hydrogenation of α,β-unsaturated citral molecule forms the basis of certain reactions carried out in perfumery industry [26].

4.9. Oxidation of toluene

Mixed metal oxide nanoparticles containing noble metals (Au, Ag, Pt, etc.) can be used in oxidative catalysis of toluene. In a recent research, a cheaper non-noble metal manganese (Mn) is reported to be used in this process (equation 7) [64]. A comparison between Mn nanocatalysts prepared via microemulsion and precipitation techniques showed that surface area of the former was much greater and hence the catalytic activity was also higher. This activity was approximately the same as that of a conventionally prepared palladium-based nanocatalyst. Therefore, a cheaper nanocatalyst with improved efficiency can be prepared using microemulsion technique:

$$CH_3 - C_6H_5 \xrightarrow{Mn} C_6H_5 - COOH \tag{7}$$

4.10. Catalytic reforming

Nanocatalysts synthesized in microemulsion medium are applicable in certain reforming process, such as bi-functional reforming of naphtha. These catalysts have greater number of active sites on their surface, thus enhancing their activity. The deactivation of such catalysts is comparatively much slow. They also increase the yield of main product, that is, toluene and benzene, and decrease the yield of unwanted by-products [92].

4.11. Oxidative coupling of methane

Oxidative coupling of methane is an effective method, which makes use of natural gas (CH_4) as an industrial feed stock. He and coworkers [93] utilized microemulsion technique to synthesize CeO_2/ZnO nanocatalysts for this purpose (Eq. (8)). They reported that the methane coupling over this specific catalyst prepared in microemulsion medium was much higher than the catalysts prepared by other techniques:

$$2\,CH_4 + O_2 \xrightarrow{CeO_2/ZnO} C_2H_4 + 2\,H_2O \tag{8}$$

4.12. Oxygen reduction reaction

Microemulsion route is also explored for the preparation of nanoparticles, which can be employed as catalyst in oxygen reduction reactions (ORRs) (most important reaction in life processes, Eq. 9). For instance, PtFe nanocatalysts supported on carbon are reported to be used in ORR. Using another metal such as Fe with Pt reduces the cost of electro catalyst. A major problem in ORR is the presence of adsorbed hydroxide [94]

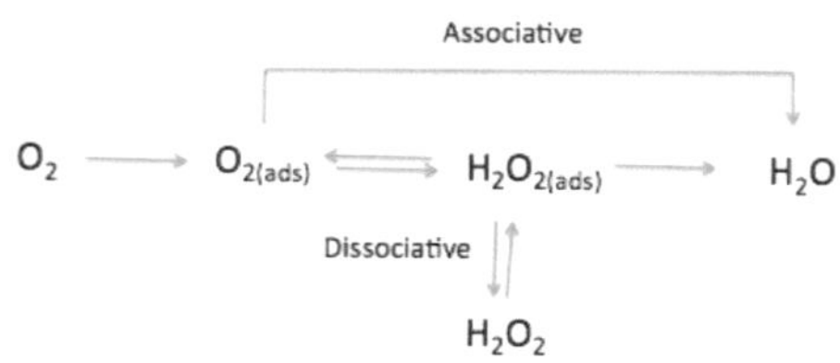

(9)

which blocks the active sites on the surface of the catalysts; this problem is overcome by using bimetallic nanocatalysts prepared in microemulsion. These catalysts can be made cathode in proton exchange membrane fuel cell (PEMFC), improving the performance of fuel cell. PEMFC in return is an important candidate for portable and vehicle applications [95].

4.13. Environmental application

The emission of air pollution into the atmosphere from automobile engines can be reduced by the use of TWCs. It can be introduced in engines as an after treatment system. Incorporation of zirconium ions in CeO_2 increases its surface area and thermal stability. Use of this catalyst enhances the activity of TWCs. However, copper can also be included in the system, which catalyzes the oxidation process of organic compounds and carbon monoxide [31, 96]. For this synthesis, microemulsion technique combined with deposition precipitation technique can be employed [31].

An important reaction in environmental protection is hydrogenation of benzene to cyclohexane [97]. Ruthenium nanocatalysts supported on alumina are applicable to catalyze the hydrogenation process (Eq. (10)). However, owing to their better tolerance towards sulfur poisoning, recycled catalysts exhibit better performance [98].

$$C_6H_6 + 3H_2 \xrightarrow{Ru/Al_2O_3} C_6H_{12} \tag{10}$$

5. Challenges in microemulsion synthesis of nanocatalysts

Although microemulsion is found to be a novel route for the preparation of nanocatalysts applicable in several fields, the recycling and separation of the organic solvents and nanocatalyst particles from the microemulsion medium is still a challenge for scientists. Traditional separation methods such as addition of a solvent, solvent evaporation, ultracentrifugation, and so on are used for this purpose. Addition of an anti-solvent (e.g. supercritical carbon dioxide, having a relatively low dielectric constant), along with a stabilizing agent such as carboxylate perfluoropolyether surfactant, being an innocuous and cheaper method, can be applied to control the stability of the catalysts effectively [11]. Making use of a photo-destructible surfactant, actuating de-stabilization of nanocatalyst particles from the microemulsion is also an interesting method for this purpose [99]. However, the shape of the nanocatalysts separated using destabilizing agents is irregular [100]. Regardless of their successful utilization in several important reactions, use of microemulsion for nanocatalyst synthesis is not much appreciated due to the lack of any industrial technique for this purpose. The cost for the synthesis of nanoparticles is relatively higher as expensive metals (Rh, Pt, etc.) are being used for the reaction. If microemulsion-based catalysts are managed to be produced at industrial scale, it is by far much better than other cheaper techniques [26]. So in the present scenario, a chemically clean and inexpensive microemulsion-synthesized nanocatalyst having desired particle size with a narrow size distribution is a real challenge [26, 101].

6. Conclusion

Microemulsion synthesis is a powerful and feasible technique for the preparation of several metal-based unimetallic, bimetallic (supported catalysts) as well as three way nanocatalysts, showing very interesting properties in certain chemical reactions. Dispersion of precursor metal particles in the nanosized droplets of microemulsion allows controlling the size, shape, morphology and size distribution of the as-prepared product depending on factors such as the initial concentration of reactants, surfactant, cosurfactant, addition of an electrolyte, and so on. The enhanced thermal stability, high catalytic activity and selectivity of the catalysts enables their use in wide range of reactions (e.g. aquathermolysis reaction, catalytic reforming, oxygen reduction reaction, photocatalysis, electrocatalysis, etc.), important in different fields such as perfumery industry, gas turbines, fuel cells, environmental protection, and so on. Despite the fact that microemulsion route synthesizes highly catalytic nanoparticles of any desired size, the process involving noble metal particles, becomes a very expensive method. This problem can be overcome by finding out an industrial way to utilize the microemulsion route for the synthesis of nanocatalysts. Moreover, the clean recovery of synthesized nanocatalysts from the reaction mixture is still a challenge.

Acknowledgements

Authors would like to thank University of the Punjab, Lahore for providing the facility of literature.

Author details

Tajamal Hussain* and Rabia Batool

*Address all correspondence to: tajamalhussain.chem@ou.edu.pk

Institute of Chemistry University of the Punjab, Lahore, Pakistan

References

[1] Hoar TP, Schulman JH. Transparent water-in-oil dispersions: the oleopathic hydro-micelle. Nature. 1943;**152**:102–103. DOI: 10.1038/152102a0

[2] Salons C, Kunieda H. Industrial Applications of Micro-emulsions. New York: Marcel Dekker; 1997.

[3] Paul Bk, Moulik SP. Microemulsions: an overview. Journal of Dispersion Science. 1997;**18**(4):301–367. DOI: 10.1080/01932699708943740.

[4] Ruckenstein E, Chi JC. Stability of microemulsions. Journal of the Chemical Society. 1975;**71**:1690–1707. DOI: 10.1039/F29757101690

[5] Schulman JH, Stoeckenius W, Prince LM. Mechanism of formation and structure of micro emulsions by electron microscopy. Journal of Physical Chemistry. 1959;**63**(10): 1677–1680. DOI: 10.1021/j150580a027

[6] Danielsson I, Lindman B. The definition of microemulsion. Colloids and Surfaces. 1981;**3**(4):391–392. DOI: 10.1016/0166-6622(81)80064-9.

[7] Eriksson S, Nylen U, Rojas S, Boutonnet M. Preparation of catalysts from microemulsions and their applications in heterogeneous catalysis. Applied Catalysis A. 2004;**265**(5): 207–219. DOI: 10.1016/j.apcata.2004.01.014

[8] Malik MA, Wani MY, Hashim MA. Microemulsion method: a novel route to synthesize organic and inorganic nanomaterials. Arabian Journal Chemistry. 2012;**5**(4):397–419. DOI: 10.1016/j.arabjc.2010.09.027

[9] Zabidi NAM. Synthesis of Nanocatalysts via Reverse Microemulsion Route for Fischer-Tropsch Reactions. In: Dr. Reza Najjar, editor. Microemulsions—An Introduction to Properties and Applications. Intech; 2012. DOI: 10.5772/37152.

[10] Lawrence MJ. Surfactant systems microemulsions and vesicles as vehicles for drug delivery. European Journal of Drug Metabolism and Pharmacokinetics. 1994;**3**:257–269. DOI: pubmed/7867669

[11] Jurek AZ, Reszczyńska, Grabowska E, Zaleska A. Nanoparticles preparation using microemulsion systems. In: Reza Najjar, editor. Microemulsions—An Introduction to Properties and Applications. Intech; 2012. DOI: 10.5772/36183

[12] Paul BK, Moulik SP. Uses and applications of microemulsions. Current Science. 2001;**80**(8):25. DOI: currsci/apr252001/990

[13] Shinoda K, Shibata Y, Lindman B. Interfacial tensions for lecithin microemulsions including the effect of surfactant and polymer addition. Langmuir. 1993;**9**(5):1254–1257. DOI: 10.1021/la00029a016

[14] Prince LM. Microemulsion Theory and Practice. New York: Academic Press; 1977. 25 p. DOI: 978-0-12-565750-1.

[15] Solans C, Kunieda H. Industrial applications of microemulsions. In: Surfactant Science Series. New York: Marcel Dekker. Inc.; 1996. p. 199. DOI: 978-0824797959

[16] Kumar P, Mittal KL. Handbook of Microemulsion Science and Technology. New York: Marcel Dekker Inc.; 1999. DOI: 9780824719791 - CAT# DK9794

[17] Mackay RA. Electrochemistry in association colloids. Colloids and Surfaces A: Physiochemical and Engineering Aspects. 1994;**82**(1):1–28. DOI: 10.1016/0927-7757(93)02610-Q

[18] Baran Jr. JR, Pope GA, Wade WH, Weerasooriya V, Yapa A. Microemulsion formation with mixed chlorinated hydrocarbons. Journal of Colloid and Interface Science. 1994;**168**(1):67–72. DOI: 10.1006/jcis.1994.1394

[19] Scameborn JF, Harwell JH. Current trends and future developments in surfactant-based separations. American Chemical Society; 2000; **740**, 1–14. DOI: 10.1021/bk-2000-0740.ch001.

[20] Attwood D. Microemulsions. In: J. Kreuter, editor. Colloidal Drug Delivery System. New York: Marcel Dekker; 1994.

[21] Das S, Jain TK, Maitra A. Inorganic-organic hybrid nanoparticles from n-octyl triethoxy silane. Journal of Colloids and Interface. 2002;**252**:82–88. DOI: 10.1006/jcis.2002.8404.

[22] El-Nokaly M, Cornell D. Microemulsion and Emulsions in Foods. In: American Chemical Society, Washington, DC. 1991. DOI: 10.1021/bk-1991-0448.fw001.

[23] Adachi M, Harada M, Shioi A, Sato Y. Extraction of amino acids to microemulsions. Journal of Physical Chemistry. 1991;**95(20)**:7925–7931. DOI: 10.1021/j100173a068.

[24] Shah DO. Micelles, microemulsions, and monolayers. In: Science and Technology, Marcel Dekker, New York. 1988, 1 p. DOI: 9780824799908-CAT# DK5583.

[25] Eastoe J, Hollamby MJ, Hudson L. Recent advances in nanoparticle synthesis with reversed micelles. Advances in Colloid and Interface Science. 2006;**128–130**:5-15. DOI: 10.1016/j.cis.2006.11.009.

[26] Boutonnet M, Lögdberg S, Svensson EE. Recent developments in the application of nanoparticles prepared from w/o microemulsions in heterogeneous catalysis. Current Opinion in Colloid & Interface Science. 2008;**13**:270–286. DOI: 10.1016/j.cocis.2007.10.001

[27] Rojas S, García-García FJ, Järas S, Martínez-Huerta MV, García Fierro FJ, Boutonnet M. Preparation of carbon supported Pt and PtRu nanoparticles from microemulsion: electrocatalysts for fuel cell applications. Applied Catalysis A: General. 2005;**285**:24–35. DOI: 10.1016/j.apcata.2005.02.005

[28] Kishida M, Ichiki K, Hanaoka T, Nagata H, Wakabayashi K. Preparation method for supported metal catalysts using w/o microemulsion: study on immobilization conditions of metal particles by hydrolysis of alkoxide. Catalysis Today. 2004;**93(95)**:205–209. DOI: 10.1016/S0920-5861(98)00216-8

[29] Hadi A, Ismail KN, Shah MNA. Effect of metals oxides loading on the modification of microstructure and phase transformation of nanocrystalline $CeZrO_2$ synthesized using water-in-oil-microemulsion. Procedia—Social and Behavioral Sciences. 2015;**195**: 2051–2060. DOI: 10.1016/j.sbspro.2015.06.227

[30] Shah MNA, Ismail KN, Jai J, Faeqah N, Hadi A. Synergistic effect of Cu-doped into $CesZrO_2$ nanocatalysts. Earth and Environmental Science. 2016;**36**. DOI: 10.1088/1755-1315/36/1/012052/pdf

[31] Solanki JN, Sengupta R, Murthy ZVP. Synthesis of copper sulphide and copper nanoparticles with microemulsion method. Solid State Sciences. 2010:**12(9)**:1560–1566. DOI: 10.1016/j.solidstatesciences.2010.06.021

[32] Inaba R, Fukahori T, Hamamoto M, Ohno T. Synthesis of nanosized TiO_2 particles in reverse micelle systems and their photocatalytic activity for degradation of toluene in gas phase. Journal of Molecular Catalysis A. 2006:**260(1-2)**:247–254. DOI: 10.1016/j.molcata.2006.06.053

[33] Granata G, Pagnanelli F, Hamane DN, Sasaki T. Effect of surfactant/water ratio and reagents' concentration on size distribution of manganese carbonate nanoparticles synthesized by microemulsion mediated route. Applied Surface Science. 2015:**331**:463–471. DOI: 10.1016/j.apsusc.2015.01.101

[34] Chandradass J, Balasubramanian M, Bae DS, Kim J, Kim KH. Effect of water to surfactant ratio (R) on the particle size of $MgAl_2O_4$ nanoparticle prepared via reverse micelle process. Journal of Alloys and Compounds. 2010:**491(1-2)**:L25–L28. DOI: 10.1016/j.jallcom.2009.10.213

[35] Virk HS. Effect of co-surfactant and water to surfactant molar ratio on CdS nanoparticles in microemulsion. Nano Trends: A Journal of Nanotechnology and Its Applications. 2010:**9(3)**:1–12.

[36] Solanki JN, Murthy ZVP. Highly monodispersed and sub-nano silver particles synthesis via microemulsion technique. Colloids Surfaces A. 2010;**359**:31. DOI: pii/S0927775710000828

[37] Solanki JN, Murthy ZVP. Controlled size silver nanoparticles synthesis with water-in-oil microemulsion method: a topical review. Industrial & Engineering Chemistry Research. 2011:**50(22)**:12311–12323. DOI: 10.1021/je201649x

[38] Zhang W, Qiao X, Chen J. Synthesis of nanosilver colloidal particles in water/oil microemulsion. Colloids Surfaces A. 2007;**29**:22. DOI: article/pii/S0927775706008545

[39] Wongwailikhit K, Horwongsakul S. The preparation of iron (III) oxide nanoparticles using W/O microemulsion. Materials Letters. 2011;**65**:2820–2822. DOI: 10.1016/j.matlet.2011.05.063

[40] Chen DH, Wu SH. Synthesis of nickel nanoparticles in water-in-oil microemulsions. Chemistry of Materials. 2000;**12(5)**:1354–1360. DOI: 10.1021/cm991167y

[41] Bagwe RP, Khilar KC. Effects of the intermicellar exchange rate and cations on the size of silver chloride nanoparticles formed in reverse micelles of AOT. Langmuir. 1997:**13(24)**:6432–6438. DOI: 10.1021/la9700681

[42] Boutonnet M, Kizling J, Stenius P. The preparation of monodisperse colloidal metal particles from microemulsion. Colloids and Surfaces. 1982;**5 (3)**: 209–225. DOI: 10.1016/0166-6622(82)80079-6

[43] Solanki JN, Ventaka Z, Murthy P. Reduction of nitro aromatic compounds over Ag/Al2O3 nanocatalyst prepared in water-in-oil microemulsion: effects of water-to-surfactant mole ratio and type of reducing agent. Industrial and Engineering Chemistry Research. 2011;**50**:7338–7344. DOI: 10.1021/ie200536q

[44] Park JC, Lee HJ, Kim JY, Park KH, Song H. Catalytic hydrogen transfer of ketones over Ni@SiO_2 yolk-shell nanocatalysts with tiny metal cores. Journal of Physical Chemistry C. 2010:**114(14)**:6381–6388. DOI: 10.1021/jp1003215

[45] Zhang W, Qiao X, Chen J. Synthesis of silver nanoparticles—effects of concerned parameters in water/oil microemulsion. Materials Science and Engineering: B. 2007;**142(1)**: 1–15. DOI: 10.1016/j.mseb.2007.06.014

[46] Malheiro AR, Varanda LC, Perez J, Villullas HM. The aerosol OT + n-butanol + n-heptane + water system:? phase behavior, structure characterization, and application to Pt70Fe30 nanoparticle synthesis. Langmuir. 2007;**23(22)**:11015–11020. DOI: 10.1021/la702146q

[47] Chiang CL, Hsu MB, Lai LB. Control of nucleation and growth of gold nanoparticles in AOT/Span80/isooctane mixed reverse micelles. Journal of Solid State Chemistry. 2004;**177**:3891–3895. DOI: 10.1016/j.jssc.2004.07.003

[48] He Y, Yang B, Cheng G, Pan H. Influence of the thermodynamic stability of microemulsion on the size of nanoparticles prepared by a coupling route of microemulsion with homogeneous precipitation. Materials Letters. 2004:**58(14)**:2019–2022. DOI: 10.1016/j.matlet.2003.12.033

[49] Lisiecki I, Pileni MP. Synthesis of well-defined and low size distribution cobalt nanocrystals: the limited influence of reverse micelles. Langmuir. 2003;**19**:9486–9489. DOI: 10.1021/la0301386

[50] Eastoe J, Cox AR. Colloids and Surfaces A: Physiochemical and Engineering Aspects. "Formation of PbS nanoclusters using reversed micelles of lead and sodium Aerosol-OT" 1995;**101**:63–76. DOI: 10.1016/0927-7757(95)03177-F

[51] Rivas J, Sánchez RD, Fondado A, Izco C, García-Bastida AJ, Garcia-Otero J, Mira J, Baldomir D, González A, Lado I, Quintela MAL, Oseroff SB. Structural and magnetic characterization of Co particles coated with Ag. Journal of Applied Physics. 1994;**76**:6564. DOI: 10.1063/1.358195

[52] López-Quintela MA, Rivas J, Blanco MC, Tojo C. Synthesis of Nanoparticles in Microemulsions. In: Nanoscale Materials. Kluwer Academic Plenum Publication, Dordrecht, Netherlands 2003.

[53] Lo ́pez-Quintela MA. Synthesis of nanomaterials in microemulsions: formation mechanisms and growth control. Current Opinion in Colloid and Interface Science. 2003;**8**: 137–144. DOI: 10.1016/S1359-0294(03)00019-0

[54] Fang J, Stokes L, Weimann J, Zhou W. Nanocrystalline bismuth synthesized via in situ polymerization–microemulsion process. Material Letters. 2000;**42**:113–120. DOI: 10.1016/S0167-577X(99)00169-X

[55] Rong M, Zhang M, Liu H, Zeng H. Synthesis of silver nanoparticles and their self-organization behavior in epoxy resin. Polymer. 1999;**40**:6169–6178. DOI: 10.1016/S0032-3861(98)00820-9

[56] Wikander K, Ekstrom H, Palmqvist AEC, Lundblad A, Holmberg K, Lindbergh G. Alternative catalysts and carbon support material for PEMFC. Fuel Cells.2006;**6(1)**:21–25. DOI: 10.1002/fuce.200500092

[57] De Moraes IR, da Silve WJ, Tronto S, Rosolen JM. Carbon fibers with cup-stacked-type structure: an advantageous support for Pt–Ru catalyst in methanol oxidation. Journal of Power Sources. 2006;**160(2)**:997–1002. DOI: 10.1016/j.jpowsour.2006.02.014

[58] Yan MC, Chen F, Zhang JL, Anpo M. Preparation of controllable crystalline titania and study on the photocatalytic properties. Journal of Physical Chemistry B. 2005;**109(18)**:8673–8678. DOI: 10.1021/jp046087i

[59] Yoon B, Wai CM. Microemulsion-templated synthesis of carbon nanotube-supported Pd and Rh nanoparticles for catalytic applications. Journal of the American Chemical Society. 2005;**127**:17174–17175. DOI: 10.1021/ja055530f

[60] Simoes M, Baranton S, Coutanceau C. Electrooxidation of sodium borohydride at Pd, Au, and $PdxAu_{1-x}$ carbon-supported nanocatalysts. Journal of Physical Chemistry C. 2009;**113(30)**:13369–13376.

[61] Zarur AJ, Ying JY. Reverse microemulsion synthesis of nanostructured complex oxides for catalytic combustion. Nature. 2000;**403**:65–67. DOI: 10.1038/47450

[62] Pour AN, Housaindokht MR, Tayyari SF, Zarkesh J. Fischer-Tropsch synthesis by nano-structured iron catalyst. Journal of Natural Gas Chemistry. 2010;**19(3)**:284–292. DOI: 10.1016/S1003-9953(09)60059-1

[63] Tavasoli A, Karimi S, Shoja M. Synthesis and application of CNT-supported MoO_3-K_2O nanocatalyst using microemulsion technique: role of nanoparticle size on catalyst activity and selectivity in higher alcohol synthesis. International Journal of Industrial Chemistry. 2013;**4**:21. DOI: 10.1186/2228-5547-4-21

[64] Li WB, Chu WB, Zhuang A, Hua J. Catalytic oxidation of toluene on Mn-containing mixed oxides prepared in reverse microemulsions. Catalysis Today. 2004;**93(95)**: 205–209. DOI: 10.1016/j.cattod.2004.06.042

[65] Fuerte A, Hernández-Alonso MD, Maira AJ, Martínez-Arias A, Fernández-García M, Conesa JC, Soria J, Munuera G. Nanosize Ti–W Mixed oxides: effect of doping level in the photocatalytic degradation of toluene using sunlight-type excitation. Journal of Catalysis. 2002;**212(1)**:1–9. DOI: 10.1006/jcat.2002.3760

[66] Wang X, Shih K, Li XY. Photocatalytic hydrogen generation from water under visible light using core/shell nano-catalysts. Water Science and Technology. 2010;**61**:9. DOI: 10.2166/wst.2010.147

[67] Mohapatra P, Mishra T, Parida KM. Effect of microemulsion composition on textural and photocatalytic activity of titania nanomaterial. Applied Catalysis A General. 2006;**319**:183–189

[68] Aman D, Zaki T, Mikhail S, Selim SA. Synthesis of a perovskite $LaNiO_3$ nanocatalyst at a low temperature using single reverse microemulsion. Catalysis Today. 2011;**164**: 209–213. DOI: 10.1016/j.cattod.2010.11.034

[69] Marella M, Tomaselli M. Synthesis of carbon nanofibers and measurements of hydrogen storage. Carbon. 2006;**44(8)**:1404–1413. DOI: 10.1016/j.carbon.2005.11.020

[70] Hyne J, Clark P, Clarke R, Koo J, Greidanus J, Tyrer J, Verona D, Aquathermolysis of heavy oils. Revista Tecnica Intevep. 1982;**2**:87–94.

[71] Chen Y, Wang Y, Wu C, Xia F. Laboratory experiments and field tests of an amphiphilic metallic chelate for catalytic aquathermolysis of heavy oil. Energy and Fuels. 2008;**22**:1502–1508. DOI: 10.1021/ef8000136

[72] Chao K, Chen Y, Li J, Zhang X, Dong B. Upgrading and vis-breaking of super-heavy oil by catalytic aquathermolysis with aromatic sulfonic copper. Fuel Processing Technology. 2012;**104**:174–180. DOI: 10.1016/j.fuproc.2012.05.010

[73] Wu Z, Zhang M, Zhao Z, Li W, Tao K. Synthesis of a Pd on Ni-B nanoparticle catalyst by the replacement reaction method for hydrodechlorination. Journal of Catalysis. 2008;**256**:323–330. DOI: 10.1016/j.jcat.2008.03.026

[74] Wallin M, Cruise N, Klement U, Palmqvist A, Skoglundh M. Preparation of Mn, Fe and Co based perovskite catalysts using microemulsions. Colloids Surface A: Physicochemical and Engineering Aspects. 2004;**238**:27–35.

[75] Giannakas AE, Ladavos AK, Pomonis PJ. Preparation, characterisation and investigation of catalytic activity for NO + CO reaction of $LaMnO_3$ and $LaFeO_3$ perovskites prepared via microemulsion method. Applied Catalysis B: Environmental. 2004;**49**:147–158. DOI: 10.1016/j.apcatb.2003.12.002

[76] Hussain T, Mujahid A, Ashraf M, Tufail F, Raza H, Shehzad K. Modified metal catalysts with enhanced surface area and their activity. NUST Journal of Engineering Sciences. 2014;**7(1)**:38–41.

[77] Yeung CM, Meunier F, Burch R, Thompsett D, Tsang SC. Comparison of new microemulsion prepared "Pt-in-Ceria" catalyst with conventional "Pt-on-Ceria" catalyst for water–gas shift reaction. Journal of Physical Chemistry B. 2006;**110(17)**:8540–8543. DOI: 10.1021/jp061083u

[78] Dry ME. High quality diesel via the Fischer–Tropsch process—a review. Journal of Chemical Technology and Biotechnology. 2001;**77(1)**:43–50. DOI: 10.1002/jctb.527.

[79] Chen W, Fan Z, Pan X, Bao X. Effect of confinement in carbon nanotubes on the activity of Fischer-Tropsch iron catalyst. Journal of American Chemical Society. 2008;**130(29)**: 9414–9419. DOI: 10.1021/ja8008192.

[80] Trepanier M, Dalai AK, Abastzoglou N. Synthesis of CNT-supported cobalt nanoparticle catalysts using a microemulsion technique: role of nanoparticle size on reducibility, activity and selectivity in Fischer–Tropsch reactions. Applied Catalysis A: General. 2010;**374**:79–86. DOI: 10.1016/j.apcata.2009.11.029

[81] Xiong LF, He T. Synthesis and characterization of ultrafine tungsten and tungsten oxide nanoparticles by a reverse microemulsion-mediated method. Chemistry of Materials. 2006;**18(9)**:2211–2218. DOI: 10.1021/cm052320t

[82] Guo L, Huang Q, Li XY, Yang S. Iron nanoparticles: synthesis and applications in surface enhanced Raman scattering and electrocatalysis. Physical Chemistry Chemical Physics. 2001;**3**:1661–1665. DOI: 10.1039/B009951L

[83] Siné G, Smida D, Limat M, Fòti G, Comninellis C. Microemulsion synthesized Pt/Ru/Sn nanoparticles on BDD for alcohol electro-oxidation. Journal of the Electrochemical Society. 2007;**1524**:170–174. DOI: 10.1149/1.2400602

[84] Raghuveer V, Ferreira PJ, Manthiram A. Comparison of Pd–Co–Au electrocatalysts prepared by conventional borohydride and microemulsion methods for oxygen reduction in fuel cells. Electrochemistry Communications. 2006;**8(5)**:807–814. DOI: 10.1016/j.elecom.2006.03.022

[85] Rabia J, Tajamal H, Saliha S, Shahid AM. Magnetic field effects on the microstructural variation of electrodeposited nickel film. Journal of Materials Science and Engineering A. 2011;**1**:481–487.

[86] Ritzkopf I, Vukojevic S, Weidenthaler C, Grunwaldt JD, Schüth F. Decreased CO production in methanol steam reforming over Cu/ZrO_2 catalysts prepared by the microemulsion technique. Applied Catalysis A General 2006;**302(2)**:215–223. DOI: 10.1016/j.apcata.2006.01.014

[87] Takenaka S, Huri KH, Matsune H, Kishida M. Control of selectivity based on the diffusion rates of the reactants in the oxidation of mixed hydrocarbons with molecular oxygen over silica-coated Pt catalysts. Chemistry Letters 2005;**34(12)**:1594–1595.

[88] Takenaka S, Umebayashi H, Tanabe E, Matsune H, Kishida M. Specific performance of silica-coated Ni catalysts for the partial oxidation of methane to synthesis gas. Journal of Catalysis 2007;245(2):392–400. DOI: 10.1016/j.jcat.2006.11.005

[89] Xu S, Zhao R, Wang XL. Highly coking resistant and stable Ni/Al_2O_3 catalysts prepared by W/O microemulsion for partial oxidation of methane. Fuel Processing Technology 2004;**86(2)**:123–33. DOI: 10.1016/j.fuproc.2003.12.013

[90] Santos L, Oliveira CHF, Moraes IR, Ticianelli EA. Oxygen reduction reaction in acid medium on Pt–Ni/C prepared by a microemulsion method. Journal of Electroanalytical Chemistry 2006;**596(2)**:141–148. DOI: 10.1016/j.jelechem.2006.07.033

[91] Meric P, Yu KMK, Tsang SC. Micelle-hosted palladium nanoparticles catalyze citral molecule hydrogenation in supercritical carbon dioxide. Langmuir. 2004;**20 (20)**:8537–8545. DOI: 10.1021/la049549s

[92] Asrami MR, Nejati B, Tavasoli A, Karimi A. Performance enhancement of Pt-Re/ Al_2O_3 naphtha reforming nanocatalysts using microemulsion. Petroleum & Coal. 2016;**58(1)**:56–65.

[93] He Y, Yang BL, Cheng GX. Controlled synthesis of CeO_2 nano-particles from the coupling route of homogeneous precipitation with microemulsion. Materials Letters. 2003;**57**:1880–1884. DOI: 10.1016/S0167-577X(02)01093-5

[94] Paulus UA, Wokaun A, Scherer GG, Schmidet TJ, Stamenkovic V, Radmilovic V, Ross PN. Oxygen reduction on carbon-supported Pt-Ni and Pt-Co alloy catalysts. The Journal of Physical Chemistry B. 2002;**106(16)**:4181–4191. DOI: 10.1021/jp013442l

[95] Malherio AR, Perez J, Villullas HM. Well-alloyed PtFeOC nanocatalysts of controlled composition and same particle size: oxygen reduction and methanol tolerance. Journal of The Electrochemical Society. 2009;**156(1)**:51–58. DOI: 10.1149/1.3006082

[96] Escribano VS, Martinez CH, Lopez EF, Amores JMG, Busca G. Characterization of a ceria-zirconia-supported Cu oxides catalyst: An FT-IR study on the catalytic oxidation of propylene. Catalysis Communications. 2009;**10(6)**:861–864. DOI: 10.1016/j.catcom.2008.12.024.

[97] Zonetti PC, Landers R, Jose A, Cobo G. Thermal treatment effects on the Ru/CeO_2 catalysts performance for partial hydrogenation of benzene. Applied Surface Science. 2008;**254(21)**:6849–6853. DOI: 10.1016/j.apsusc.2008.04.097.

[98] Nandanwar SU, Chakrabortry M, Mukhopadhyay S, Shenoy KT. Benzene hydrogenation over highly active monodisperse Ru/c-Al_2O_3 nanocatalyst synthesized by (w/o) reverse microemulsion. Progress in Reation Kinetics and Mechanism. 2012;**108(2)**: 473–489. DOI: 10.1007/s11144-012-0526-1.

[99] Eastoe J. Photo-destructible surfactants in microemulsions. Progress in Colloid and Polymer Science. 2006;**133**:106–110. DOI: 10.1007/2882_047.

[100] Vesperinas A, Eastoe J, Jackson S, Wyatt P. (2007). Light-induced flocculation of gold nanoparticles. Chemical Communications. 2007; 3912–3914. DOI: 10.1039/B710502A.

[101] Salabat A, Eastoe J, Vesperinas A, Tabor RF, Mutch KJ. Photorecovery of nanoparticles from an organic solvent. Langmuir. 2008;**24(5)**:1829–1832. DOI: 10.1021/la703145n.

Section 3

Theoretical and Physical Parameters Predicting Microemulsion Properties

Chapter 5

Investigation of Liquid-Phase Inhomogeneity on the Nanometer Scale Using Spin-Polarized Paramagnetic Probes

Valery F. Tarasov and Malcolm D.E. Forbes

Additional information is available at the end of the chapter

http://dx.doi.org/10.5772/67463

Abstract

The concept, basic physics, and experimental details of time-resolved electron paramagnetic resonance (TREPR) spectroscopy for the study of spin-correlated radical pairs (SCRPs) in heterogeneous media are presented and discussed. The delicate interplay between electron spin wave function evolution (governed by magnetic interactions such as the isotropic electron spin-spin exchange interaction and the electron-nuclear hyperfine interaction) and diffusion (governed by the size and microviscosity of the medium) provides a mechanism for assessing molecular mobility in confined spaces on the nanoscale (e.g., micelles, vesicles, and microemulsions). Experimental examples from micellar SCRPs are used to highlight the dominant features of the TREPR under different degrees of confinement and microviscosity, and spectral simulation methods are described to show how molecular mobility can be quantified.

Keywords: electron paramagnetic resonance, spin-correlated radical pairs, micelles, diffusion, microviscosity, confinement

1. Introduction

Electron paramagnetic resonance (EPR) spectroscopy is a fundamental method for the investigation of paramagnetic particles in liquid and solid solutions, in crystal and semiconductor substances, and in biological and physiological systems. Examples of such paramagnetic species include organic free radicals, transition metals such as Fe, Cu, Co, Ni, Mn, and their complexes, particularly dinitrosyl-iron complexes with thiol-containing ligands [1], excited molecular states (especially excited molecular triplets), quantum dots, and crystal defects. The EPR spectra of these systems provide valuable structural and dynamic information through spectral line shapes, g-factors, electron-nuclear hyperfine interactions, and electron spin relaxation processes. The electron and nuclear spin relaxation parameters depend on interactions with their nearest paramagnetic neighbors and also on their molecular mobility. In this sense, all of the paramagnetic species mentioned above can be considered as useful probes of molecular motion and intermolecular interactions. However, extraction of the desired information from the EPR spectra can be time-consuming and this is not a simple computational task.

Nitroxide free radicals occupy an exceptional place among these species because they are minimally invasive and chemically stable. Their facile synthesis allows for the creation of a whole family of probes with different chemical structures and properties. Also, their EPR spectra are relatively simple, with only three resonant EPR transitions in most cases. The quantitative treatment of their spectra in isotropic media is based on a well-established and comprehensive theoretical analysis that is well documented in the literature [2–5]. For more complex situations involving the appearances of spectral anisotropies, or in the limit of slow motion, easily accessible and efficient computer programs have evolved for carrying out numerical simulations [6, 7] of the spectra. It is no surprise, therefore, that when the term "spin probe method" or "spin labeling" is used, it almost always refers first to the analysis of the EPR spectra of nitroxide radicals.

The nitroxide spin probe/label method has found wide applications in chemistry, biochemistry, and physics. Modern development of the method is deeply connected to recent advancements in EPR instrumentation. Examples include an increase in the microwave frequency [8, 9] up to 215–240 GHz (very high frequency EPR, VHF/EPR), application of pulsed EPR spectrometers [9, 10], as well as advances in targeted spin labeling (site-directed spin labeling, SDSL [11–13]). New chemical synthesis methods have resulted in novel probe structures designed, for example, to measure the polarity of the microenvironment surrounding the probe [14], the concentration of oxygen [15], and other properties in both homogeneous and heterogeneous systems.

There also exists a wide range of short-lived hydrocarbon radicals, typically transient species, that are observed as a result of photochemically or thermally initiated radical reactions. For some time, detectable EPR spectra of these intermediates were used only to verify or refute a particular mechanism of a chemical reaction [16, 17]. However, in 1963, Fessenden and Schuler reported highly unusual EPR spectra of hydrogen and deuterium atoms. Specifically, the low-field components of the hydrogen doublet and deuterium triplet were emissive, while the high-field components were absorptive [18]. These spectra were acquired at low temperatures during pulse radiolysis of liquid methane and deuteromethane. This spin-polarization phenomenon would have remained a curiosity, at the time attributed to an unexplained spin relaxation effect, if not for the discovery of the chemically induced dynamic nuclear polarization (CIDNP) a few years later. The CIDNP effect is defined as the creation of nonequilibrium populations of the nuclear spin states in the diamagnetic products of radical reactions involving recombination and disproportionation. The discovery of CIDNP in the 1960s [19, 20] stimulated deeper insights into the manifestations of the well-known rule of the electron spin-state conservation in chemical reactions, initially put forward by Wigner and Witmer [21]. By the end of the 1960s Closs, and independently, Kaptein and Oosterhoff, developed the so-called CKO model of CIDNP, which has been described in detail in several monographs (see for instance [22, 23]) and reviews [24–26]. This model allowed for a qualitative interpretation of all significant details of CIDNP but failed to provide much quantitative information. The main drawback of the CKO model is an assumption of unrealistically long lifetimes for the radical pairs involved and a total disregard of the exchange spin-spin interaction between the unpaired electrons of the radical partners.

Consideration of the mutual diffusion of radicals as a continuum process under the modified Collins and Kimball boundary conditions [22, 27], or as Noyes diffusion flights (the so-called method of successive encounters [28]), removed these shortcomings and allowed researchers to formulate a model for liquid-phase spin-selective diffusion-controlled radical reactions—the radical pair mechanism (RPM) of CIDNP. In both the RPM and CKO models, the essence of CIDNP remains intact: it is based on the difference in the reactivity of singlet and triplet radical pairs.

The intensity of the spin-polarized NMR signal depends on the configuration of the nuclear spins in the radicals and on their coefficient of mutual diffusion. Therefore, at least in principle, the CIDNP magnitude could be used as a measurable parameter allowing for the extraction of information about the translation diffusion of radicals, for example, by comparing its value with the computed one. However, the RPM model is too idealized. The radicals are considered to be spherical particles, whose reactivity does not depend on orientation. The model operates in such terms as the thickness of a reaction layer in the reaction sphere or uses parameters in the Collins and Kimball boundary condition, the physical meanings of which are far from transparent. Additionally, it can operate in terms of distance-dependent reactivity of the radical pairs in the singlet electron spin state, with unknown or vague numerical values for the corresponding parameters. As a consequence, these circumstances make the quantitative conclusions about the molecular dynamics of radicals from the analysis of CIDNP data quite unreliable.

Contrary to CIDNP, the spin selectivity of radical recombination does not play such a crucial role in the generation of CIDEP. Nevertheless, nuclear spin-dependent singlet-triplet mixing is still a necessary step for the observation of CIDEP via the RPM. However, instead of a reaction taking place from a specific electron spin state, the relative phase shifts of the electron spin singlet and triplet states induced by the distance-dependent electron spin exchange interaction become a crucial factor for the development of CIDEP [29, 30]. Not all such repetitive encounters are efficient in the creation of CIDEP. The pair must not spend too long time at distances where the spin's exchange interaction is zero. Also, this time must be comparable with the difference in the Larmor frequencies of the radicals. On the other hand, the lifetime of the contact pairs must not be too short, that is, there should be enough time for efficient dephasing between the multiplicative electron spin states of a pair. Therefore, CIDEP created by the RPM, ST_0RPM in the particular case of the high-field approximation, is observed in radicals that became involved in secondary encounters after rather long (several tens of Å) diffusional journeys [29–34].

As in the case of other mechanisms of CIDEP such as the triplet mechanism (TM) [34], the electron spin polarization transfer (ESPT) [35, 36], and the radical triplet pair mechanism (RTPM) [37], the observed EPR signals, usually acquired in a time-resolved experiment, belong to "free" radicals and are characterized by the same line widths and resonance frequencies as if the radicals were in the Boltzmann equilibrium state (nonpolarized). But the molecular mobility information potentially extractable from these parameters can be obtained using the nitroxide spin probe method in a more efficient and less expensive way. On the other hand, the polarization patterns observed in TREPR experiments carry additional information about the interaction of radicals with each other. Due to the length of the diffusional trajectories, this information is much less local and therefore may be of significance to researchers working in heterogeneous media on the nanoscale (micelles, vesicles, and microemulsions).

This unique property of CIDEP via the RPM is manifested most clearly when the diffusion of the radicals is limited by a physical boundary, so that all the diffusional trajectories of the radical pair satisfy the conditions listed above. Under such conditions, each individual spectroscopic line in each radical's TREPR spectrum is split into two lines of opposite phase and gives rise to what Shushin [38], in 1991, dubbed as the "Anti-Phase Splitting," (APS). The corresponding model of CIDEP for such cases is called the spin-correlated radical pair mechanism or SCRPM [39]. In this chapter, we will revisit several systems involving radical pair confinement with restricted diffusion in the context of some unusual experimental observations. Our goal is to show that SCRPs are very valuable spin probes for self-organizing molecular systems such as micelles, vesicles, microemulsions, and potentially for any medium with the characteristic sizes of inhomogeneity on the nanometer scale.

2. Features of TREPR spectroscopy: spin-polarized "free" radicals

The TREPR technique uses continuous wave (cw) microwave excitation in the same way as commercial steady-state EPR (SSEPR) spectrometers: The TREPR transitions are detected by sweeping an external magnetic field B through each resonance at a constant microwave frequency ω_0. The TREPR operates in so-called direct detection mode (**Figure 1**) in the sense that the 100-KHz field modulation normally employed in SSEPR is disabled. **Figure 1** shows

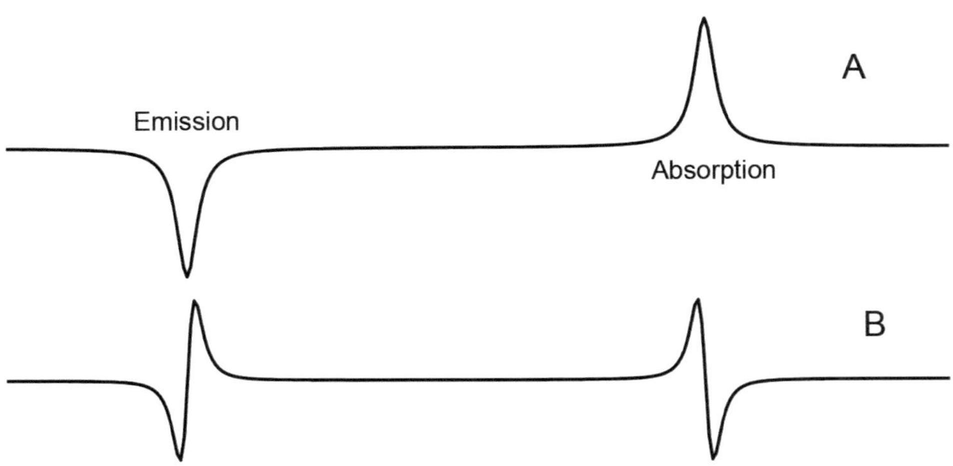

Figure 1. (A) EPR doublet signal acquired in "direct detection" mode (TREPR). (B) The same signal acquired with magnetic field modulation (SSEPR). Note that the first derivative lines in trace B have opposite phases.

the comparison of an EPR signal acquired in the "direct detection" mode (trace A, TREPR) with a signal from the same radical species detected with field modulation (trace B, SSEPR). Trace B, which shows the first derivative line shape, is exactly the signal observed by Fessenden and Schuller [18] in their discovery of electron spin polarization. Note the opposite phases of the low- and high-field transitions in both traces.

Detection of TREPR signals in the direct detection mode results in a significant loss in sensitivity compared to SSEPR. But this shortcoming is compensated efficiently by the gain in time response, since the TREPR signal is coming directly from the spectrometer's microwave bridge preamplifier circuit and is sampled electronically on a short timescale after its creation. The time response at X-band is generally limited by the resonator quality factor to about 50 ns. Higher-frequency W-band spectrometers (95 GHz) allow for much higher time resolution (down to 2.5 ns) [37]. Unfortunately, technical problems, such as delivering light to the sample and the difficulty of using flow systems, significantly narrow the applicability of W-band spectrometers.

The Lorentzian line shape of the resonance signals in SSESR experiments is ensured in liquid solutions via the steady state Bloch equations, that is, under conditions of the absence of saturation, the time interval between the creation of magnetization and its detection t_{obs} should be much longer than the relaxation time T_2 or $t_{obs}/T_2 \gg 1$. Typical transverse relaxation times of organic radicals range from tens of nanoseconds to several microseconds. This means that in many cases, the condition $t_{obs}/T_2 \gg 1$ is unfulfilled. It is worthwhile to consider in this case, what happens to the resonance line shape.

Figure 2 shows TREPR spectra of geminate radical pairs (**Scheme 1**) resulting from photo-induced Norrish type-I cleavage (Nd-YAG, $\lambda = 355$ nm) of 2,2-dimethoxy-1,2-diphenyl ethane-1-one (trade name: Irgacure®651) in aqueous SDS micellar solutions at room temperature [40, 41]. The TREPR spectra in **Figure 2** were acquired at the outlet of a microwave bridge-wide bandwidth preamplifier (12 MHz gate), using a LeCroy digital oscilloscope at $t_{obs} = 260$ ns (A), $= 1.12$ μs (B), and $= 3.6$ μs (C)

The spectra in **Figure 2** are strongly polarized by the TM. The most striking feature of the spectra is the strong dependence of the width of the spectral lines on the observation time t_{obs}. The line width at the longest observation time $t_{obs} = 3.6$ μs (trace C) corresponds to $T_2 = 1.2$ μs for the $\alpha\alpha$-dimethoxybenzyl radical. The condition $t_{obs}/T_2 \gg 1$ is not, strictly speaking, fulfilled even for $t_{obs} = 3.6$ μs and far from being fulfilled for the other two cases: $t_{obs} = 260$ ns and $t_{obs} =$ 1.12 μs. Thus, one can suggest that the additional line broadening is either caused by modulation of the inter-radical electron spin interactions due to the molecular diffusional motion of radicals or some inherent properties of the EPR spectra observed under conditions of the violation of the inequality $t_{obs}/T_2 \gg 1$ or both. At the moment, we do not need a detailed discussion of this problem; it will be quite enough to show how the resonance line appears in the TREPR experiment at $t_{obs}/T_2 \leq 1$. A similar analysis has been performed already using a set of modified Bloch equations [42, 43].

In a frame rotating with the angular velocity ω_0 around the direction of the magnetic field of spectrometer B, the spin and its interaction with the resonance microwave field are represented

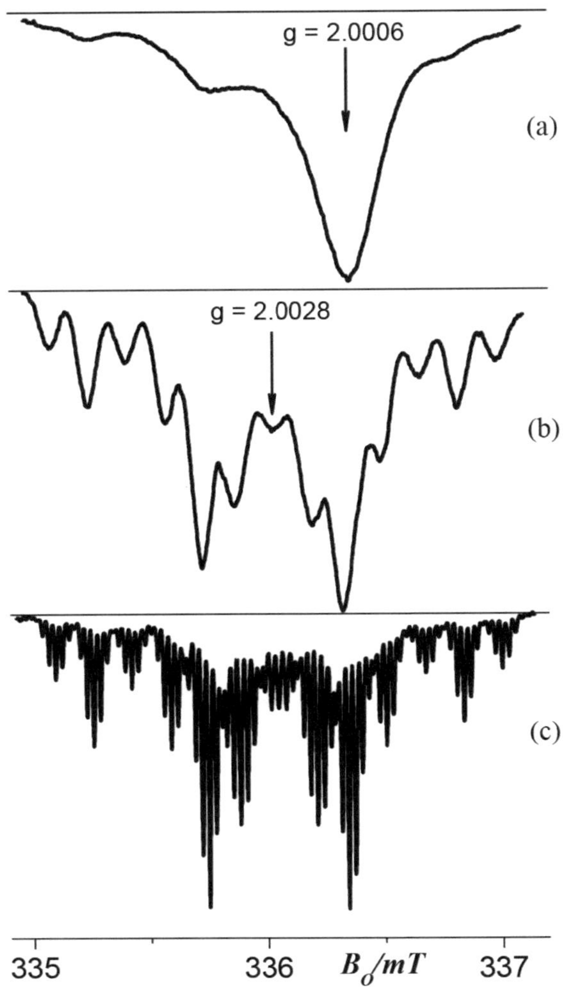

Figure 2. TREPR spectra at t_{obs} = 260 ns (a), = 1.12 μs (b), and = 3.6 μs (c) of geminate radical pairs resulting from photodissociation of Irgacure®651 (**Scheme 1**) in micellar aqueous solutions of SDS at room temperature [40, 41].

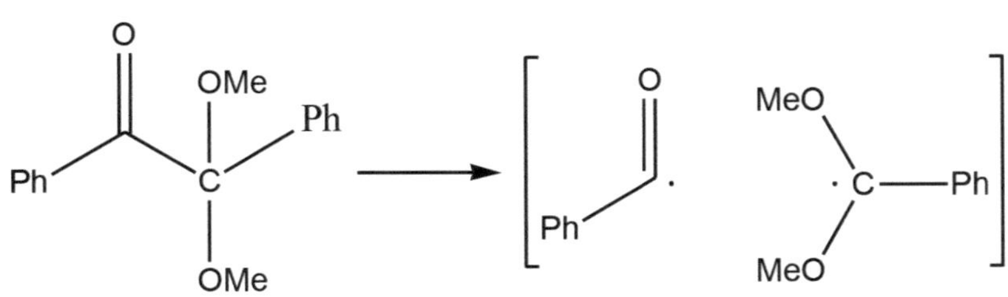

Scheme 1. Photoexcitation (355 nm) of 2,2-dimethoxy-1,2-diphenyl ethane-1-one (Irgacure®651) resulting in Norrish type-I cleavage. This reaction is very fast and forms a geminate triplet pair of benzoyl (g = 2.0006) and $\alpha\alpha$-dimethoxybenzyl (g = 2.0028) radicals.

http://dx.doi.org/10.5772/67463

by the time-independent spin operators $\hat{H}$ and $\hat{H}_{\mu w}$ respectively. Then, the corresponding Liouville equation (LE) for the spin-state vector $\boldsymbol{\rho}$ reads:

$$\frac{d\boldsymbol{\rho}}{dt} = \{-i(\hat{\hat{H}} + \hat{\hat{H}}_{\mu w}) + \hat{\hat{R}}\} \cdot \boldsymbol{\rho} \tag{1}$$

Super operators $\hat{\hat{H}}$ and $\hat{\hat{H}}_{\mu w}$ in the Liouville spin space are associated with the commutators in the Hilbert spin space $[\hat{H}, \hat{\rho}] \Rightarrow \hat{\hat{H}} \cdot \boldsymbol{\rho}$, $[\hat{H}_{\mu w}, \hat{\rho}] \Rightarrow \hat{\hat{H}}_{\mu w} \cdot \boldsymbol{\rho}$, respectively, and the state vector $\boldsymbol{\rho}$ is comprised of elements of the spin density operator $\hat{\rho}$. $\hat{\hat{R}}$ is the relaxation super operator.

This situation is commonly encountered when the particular mechanism of paramagnetic relaxation is unknown or when several mechanisms operate simultaneously. In our work, we use the phenomenological relaxation superoperator—the matrix representation of which in the eigenbasis of $\hat{H}$ and in the high-field approximation limit is given by Eq. (2):

$$\hat{\hat{R}} = \begin{pmatrix} -k_1 e^{\beta} & 0 & 0 & k_1 e^{-\beta} \\ 0 & -k_2 & 0 & 0 \\ 0 & 0 & -k_2 & 0 \\ k_1 e^{\beta} & 0 & 0 & -k_1 e^{-\beta} \end{pmatrix} \tag{2}$$

where $k_1 = 1/2T_1$ is assumed to be the same for all pairs of the electron-nuclear spin levels. We also suggest that all the off-diagonal elements of the density matrix decay with the same rate constant $k_2 = 1/T_2$. The relaxation super operator (2) ensures the Boltzmann equilibrium of the spin ½ at $t \rightarrow \infty$ if k_2 and k_1 are real and positive. Since we solve Eq. (1) in rotating frame of reference, then it is quite natural to wonder why do we do believe that in the rotating frame the spin ensemble relaxes to the **B** direction instead of the effective magnetic field in the rotating frame. See Redfield [44] for the discussion of this problem.

Eqs. (1)–(2) are applicable to any time of observation (t_{obs}) and lead to a Lorentzian shape, Eq. (3), of the resonance line if $t_{obs} \gg T_2$ and $\omega_1 << (T_1, T_2)^{-1/2}$,

$$I(\omega) = \frac{1}{\pi} \frac{T_2^{-1}}{T_2^{-2} + (\omega - \omega_r - \Delta_r)^2} = \frac{T_2}{\pi} \frac{1}{1 + (\omega - \omega_r - \Delta_r)^2 T_2^2} \tag{3}$$

where ω_r and Δ_r are the resonance frequency and the spectral shift induced by relaxation, respectively.

Figure 3 shows the TREPR spectra of an electron spin ½ being initially in the spin state $|\beta\rangle$ or $\hat{\rho_0} = \begin{pmatrix} 0 & 0 \\ 0 & 1 \end{pmatrix}$. For all calculations, $B_1 = 0.01$ G, $T_1 = 5$ μs, $T_2 = 0.25$ μs (**Figure 3A** and **3B**), and 80 ns (**Figure 3C** and **3D**). $t_{obs} = 0.2$ μs (A, B and C), and $t_{obs} = 25$ μs (Boltzmann population, Trace D). Traces A and C in **Figure 3** demonstrate what happens when the *cw* TREPR spectra are acquired at such t_{obs} that the inequality $t_{obs}/T_2 \gg 1$ is not fulfilled. Side wiggles cause an extra broadening [43] of the resonance lines. Comparing the A and C spectra in **Figure 3** suggests that when t_{obs} is comparable with T_2, the width of the resonance line is defined

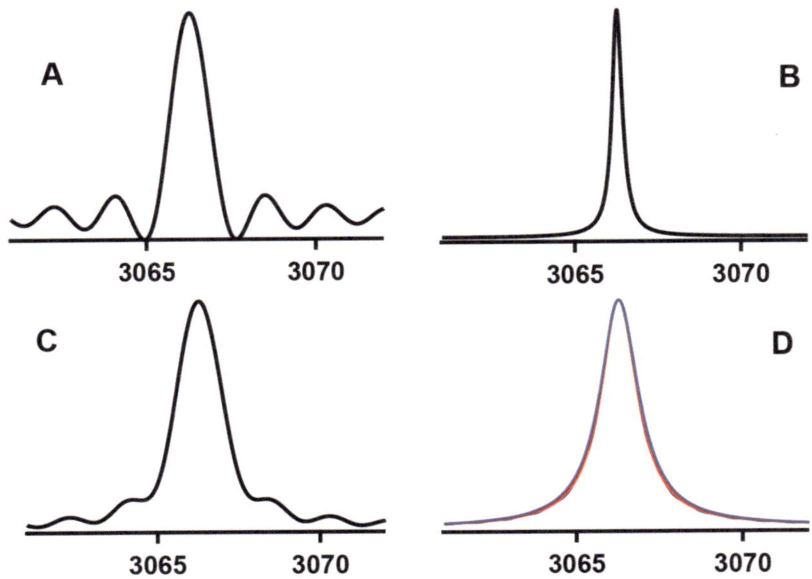

Figure 3. Calculated TREPR spectra using Eqs. 1, 2 for $B_1 = 0.01$ G and $T_1 = 5$ μs. Trace A: t_{obs} (= 0.2 μs)/T_2 (= 0.25 μs) = 0.8; Trace B: t_{obs} (= 2 μs)/T_2 (= 0.25 μs) = 8; Trace C: t_{obs} (= 0.2 μs)/T_2 (= 0.08 μs) = 2.5; Trace D (Boltzmann equilibrium): t_{obs} (= 25.6 μs)/T_2 (= 0.08 μs) = 320.

practically entirely by t_{obs} and does not depend on the transversal relaxation time T_2. A comparison of traces B and D in **Figure 3** shows that shorter the relaxation time T_2, the longer the time interval when the shape of the resonance line does not depend on t_{obs}. For instance, at $T_2 = 80$ ns, the shape of the resonance line only slightly differs from that for a radical at Boltzmann equilibrium (trace D), even at $t_{obs} = 400$ ns.

3. The SCRP mechanism and its manifestation in the TREPR spectra

In the previous paragraph, we have shown that the width of the EPR resonance line, a key parameter of the SSEPR spin probe method, is not suitable for the TREPR spin-polarized method because it depends on t_{obs}. Even when it would seem that the line width does not depend upon t_{obs}, it still remains an unreliable measurable parameter, because the shape of the TREPR resonance line is slightly different from Lorentzian, appearing slightly narrower than what would be observed for the same radical resonance line at Boltzmann equilibrium (**Figure 3D**). The polarization patterns inherent to such mechanisms of CIDEP, as ESPT, RTPM, and RPM contain very valuable information concerning the multiplicity of radical pairs, or radical–triplet pairs, the distribution of electron–nuclear spin populations in the excited spin-not-equilibrated triplet, or the signs of the exchange interactions. However, this is mostly information about the radical pair itself. Extraction of information about molecular motion from these experiments, that is, using the polarization patterns of free radicals observed in TREPR experiments, and their time evolution, is accompanied by many theoretical and computational difficulties, see for instance [42, 43, 45, 46]. This has motivated us to obtain further insight into the analysis of the CIDEP patterns of radical pairs experiencing restricted translation diffusion.

3.1. History: the quasi-static approximation (QSA)

For the better part of the past three decades, SCRPs have been a topic of keen interest in the field known as "spin chemistry." Many structurally distinct types of SCRPs have been studied by the TREPR [47–50] and NMR [51–53] spectroscopy, including resonance microwave field effects on nuclear spin polarization [54], on chemical product reaction yields [55–60], and on internal magnetic field effects [61, 62]. It is important to note that SCRPs represent a true case of entangled spin states. For this reason, SCRPs find themselves at the forefront of quantum computing research [63–66] and of high relevance to the more general field of spintronics [67].

In the EPR spectroscopy community, the term "SCRP" was put into use in 1987 by Buckley et al. [68]. The necessity of a specific term was justified by the observation of specific line shapes in the TREPR spectra resulting from photochemical reactions of acetone with isopropanol in liquid solutions: each resonance line of the acetone ketyl radical was split into two oppositely polarized components exhibiting an emission/absorption (E/A) pattern. In 1989, the nomenclature for this spectral pattern was abbreviated by Shushin [38] as the APS. It is worthwhile to note that there were a few earlier publications [69–72] that reported APS-like spectral patterns, but these reports either suggested no interpretation or interpretations that were unlikely to be correct.

An example of the APS feature in a TREPR spectrum is given in **Figure 4**, where spectrum A belongs to a radical pair (**Scheme 2**) generated by the laser flash photolysis of ^{13}C-labeled in the carbonyl group 2,2-dimethyl-desoxy-benzoin (DMDB) in an aqueous micellar solution of SDS at room temperature.

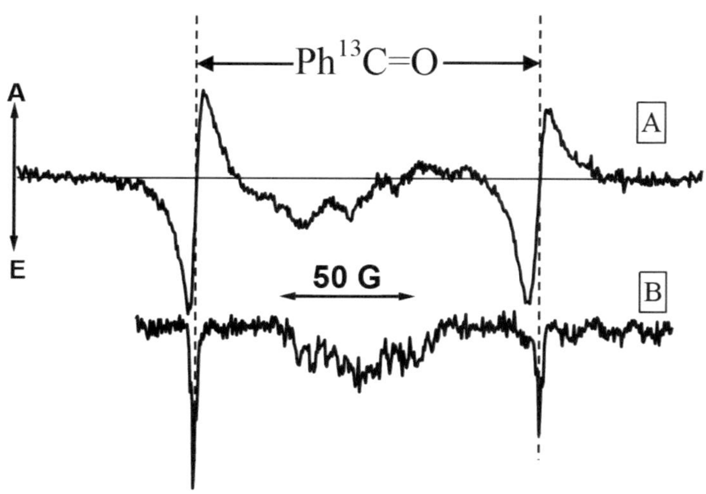

Figure 4. The TREPR spectra of radical pairs comprised of ^{13}C-benzoyl and cumyl radicals resulting from the photochemical reaction shown in **Scheme 2** via laser photolysis flash (λ = 308 nm) at room temperature in (A) aqueous SDS micellar solutions and (B) in viscous ethanol:cyclohexanol = 1:4 mixture. In both spectra, t_{obs} = 360 ns.

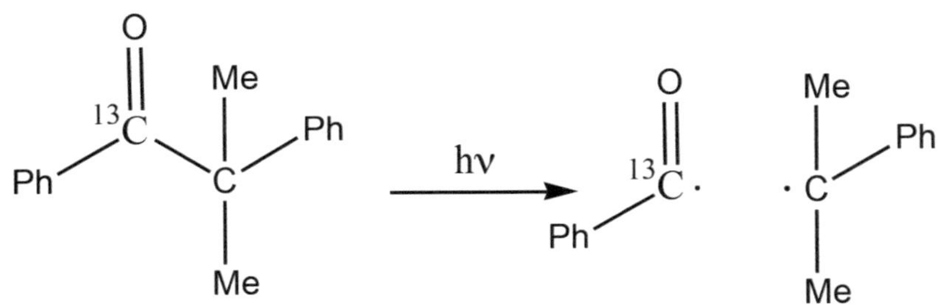

Scheme 2. Photoexcitation (λ = 308 nm) of 2,2-dimethyl-1,2-diphenylethane-1-[^{13}C] on (DMDB) resulting in a geminate triplet radical pair consisting of [^{13}C]benzoyl (g = 20006) and cumyl (g = 2.0028) radicals. See Ref. [67] for details.

Spectrum B in **Figure 4** belongs to the same radical pair acquired at the same t_{obs} = 360 ns and under the same experimental conditions (temperature, microwave power, etc.) except that spectrum B was obtained for laser photolysis in a homogeneous mixture of solvents ethanol: cyclohexanol = 1:4, which has a similar viscosity to that in the SDS micellar core. Spectra A and B in **Figure 4** are both composed of doublets from the ^{13}C-benzoyl radical and multiplet signals in the center from the cumyl radical. Each component of the benzoyl doublet is E/A split in **Figure 4A**, while in **Figure 4B**, the same doublet is TM polarized and does not contain any additional splitting. The spectrum of the cumyl radical shows the same difference in spectral features between micelles and free solution, although not quite so evidently due to the complexity of the spectrum.

Another example of a TREPR spectrum exhibiting the APS is shown in **Figure 5**. As before, the spectra in traces A and B arise from the same radical pair (**Scheme 3**) with only one difference: while trace A belongs to the radical pair generated by photoreduction of benzophenone in SDS micelles, that is, located inside the micellar phase, trace B represents the same radical pair observed in liquid solution, where they move freely. The CIDEP of the radicals in trace B is

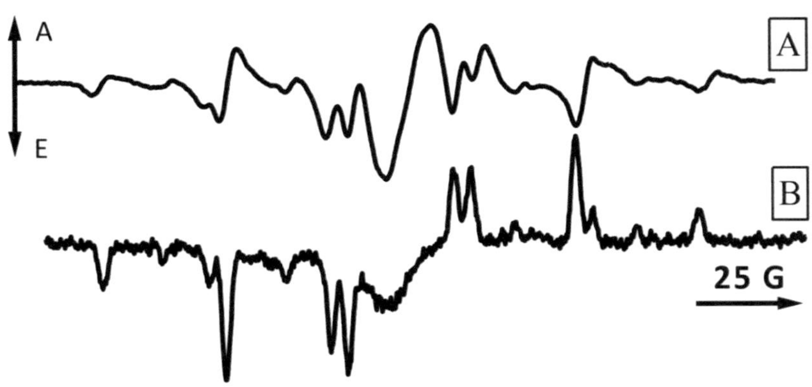

Figure 5. TREPR (t_{obs} = 500 ns) spectra of geminate pairs Alk· and BPH· radicals produced in photochemical (λ= 308 nm) oxidation of SDS detergent by the electronically excited triplet state of benzophenone (0.4 mM) in micellar aqueous solution of SDS (10 mM) (A) and benzophenone (0.4 mM) in ACN/H2O = 50:50 solution of SDS which does not contain micelles.

http://dx.doi.org/10.5772/67463

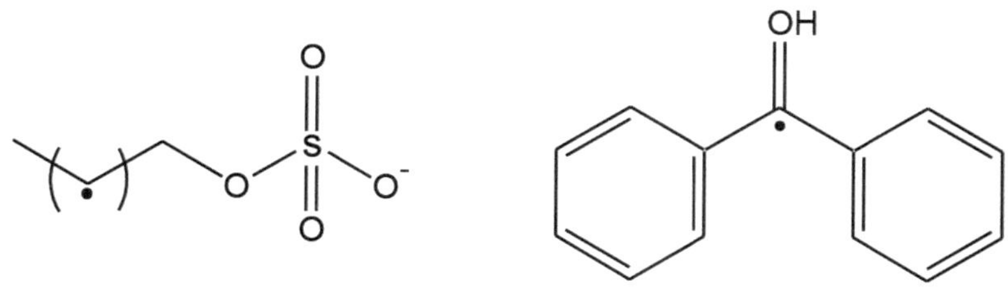

Scheme 3. Chemical structure of Alk· and BPH· radicals produced in the photochemical oxidation of SDS detergent by the electronically excited triplet state of benzophenone.

mostly due to the RPM. There is a noticeable contribution from the TM polarization, but there is no APS. Neither the TM nor the RPM polarization patterns are observed in trace A, where diffusion is restricted. The results presented in **Figures 4** and **5** are intriguing as they represent a measurable physical change that is imposed only by a boundary condition on radical diffusion. This suggests that SCRPs might be used as "polarized spin probes" to investigate molecular and spin dynamics in inhomogeneous media, even in a qualitative fashion, that is, as a method of sensing restricted mobility.

Important and pioneering theoretical work by Buckley et al. [68], Closs et al. [73] was brought forward in the 1980s to explain the APS phenomenon, and it was based on two central ideas:

1. In a chemical reaction resulting in a geminate radical pair, the population of the SCRP electron-nuclear spin states proceeds so quickly that the electron spin state of the precursor remains intact, irrespective of the magnitude of any inter-radical spin-spin interactions operating during the process of radical pair creation.

2. The presence of inter-radical electron spin exchange and dipolar magnetic interactions leaves an inherent double degeneracy of the EPR transitions in a pair of magnetically non-equivalent electron spins, resulting in an additional splitting of the EPR spectral lines similar to that what is going on in the SSEPR spectroscopy of stable nitroxide biradicals [74].

But despite the full consent in the two fundamental principles, understanding how this inter-radical interaction is manifested in the TREPR spectroscopy is completely different.

According to [68], the spectrum consists of contributions from immobilized SCRPs, with the radical partners diffused apart and separated by whatever inter-radical distances $r(t_{obs})$ have been attained at the moment of observation t_{obs}. The inter-radical electron spin exchange interaction depends on these distances and consequently on t_{obs}, that is, $J_{ex} = J\left(r(t_{obs})\right)$. It follows that the instantaneous Hamiltonian of a radical pair $H(\hat{t_{obs}})$, and its resonant magnetic fields $\omega_i(t_{obs})$ will also depend on t_{obs}. From this, we conclude that the overall spectrum was calculated as purely inhomogeneous. All possible TREPR transitions, corresponding to each instantaneous Hamiltonian, are assumed to possess the same T_2, which becomes a parameter to be varied to best fit. The intensities of the EPR resonance lines were believed to be proportional to the product of the difference in populations of corresponding electron spin eigenstates of the instantaneous Hamiltonian times the probability of the EPR transition. This is the origin of what we call the *quasi-static approximation* or QSA.

There are rather obvious disadvantages of the QSA. First, it cannot account for the generation of CIDEP due to the RPM (both ST_0 and $ST_{\pm}$), which strongly contradicts experimental observations (see **Figures 4B** and **5B**). Instead, one has to take this polarization pattern into account artificially (see **Figure 3** in Ref. [68]). Second, the TREPR spectrum at t_{obs} for the instantaneous $\hat{H}(t_{obs})$ would be a single resonance line with a particular T_2 (see Section 1 of the chapter) only if the instantaneous Hamiltonian evolved adiabatically (in the sense of energy levels) or nonadiabatically (in the sense of spin quantum numbers). In other words, a pair of acetone ketyl radicals would have to diffuse so slowly that they remained at distances r (t_{obs}), where the exchange interaction is comparable to the hyperfine interaction, for at least such a time t_r that $t_r > T_2$, which may not be realistic. Third, T_2 cannot be considered as an independent parameter because modulation of the inter-radical electron spin interactions can be the main reason for spectral line broadening (T_2). For example, see **Figures 4A** and **5A**, where the components of the APS are much broader than those in **Figures 4B** and **5B**.

The instantaneous spin Hamiltonian $\hat{H}(J)$ of a pair of radicals ***a*** and ***b*** in the high-field approximation can be written as the sum:

$$\hat{H}(J) = \hat{h}_a + \hat{h}_b + \hat{J}_{ex} \tag{4}$$

where

$$\hat{J}_{ex} = -J_0 \cdot \left(\frac{1}{2} + 2\hat{\mathbf{S}}_a \cdot \hat{\mathbf{S}}_b\right) \cdot \exp\left(-\frac{r-R}{\lambda}\right) \tag{5}$$

describes the instantaneous inter-radical electron spin-spin Heisenberg exchange interaction, R is the radius of the contact sphere (the distance of closest approach for the case of spherical radicals), J_0 is the exchange interaction at R, and the spin Hamiltonians of separated radicals $\hat{h}_a$ and $\hat{h}_b$ are given by Eq. (6).

$$\hat{h}_a = \frac{1}{2} \cdot \begin{pmatrix} \omega_a & 0 \\ 0 & -\omega_a \end{pmatrix} \otimes \begin{pmatrix} 1 & 0 \\ 0 & 1 \end{pmatrix}; \ \hat{h}_b = \begin{pmatrix} 1 & 0 \\ 0 & 1 \end{pmatrix} \otimes \frac{1}{2} \cdot \begin{pmatrix} \omega_b & 0 \\ 0 & -\omega_b \end{pmatrix} \tag{6}$$

ω_a and ω_b are the Zeeman frequencies of the electron spins in the magnetic field of the spectrometer and for a particular configuration of the magnetic nuclei $\{\chi\} = \{\chi_a\} \otimes \{\chi_b\}$ in SCRP which is defined by nuclear spin configurations $\{\chi_a\}$ and $\{\chi_b\}$ in radicals. The eigenvalues and eigenfunctions in the multiplicative electron spin basis

$$|\alpha\alpha; \chi\rangle, \ |\alpha\beta; \chi\rangle, \ |\beta\alpha; \chi\rangle, \ |\beta\beta; \chi\rangle \tag{7}$$

of the spin Hamiltonian (4) are given in **Table 1**. If $2q > 0$ and $J < 0$, the electron spin states at $-J \gg q$ (contact pair) and at $J = 0$ (separate pair) correlate as shown in **Figure 6**. When $J \neq 0$ two twice degenerated EPR transitions in the system of two non-interacted ($J = 0$) radicals ***a*** and *b* $|\psi_1\rangle \leftrightarrow |\psi_3\rangle = |\alpha\alpha\rangle \leftrightarrow |\beta\alpha\rangle$; $|\psi_4\rangle \leftrightarrow |\psi_2\rangle = |\beta\beta\rangle \leftrightarrow |\alpha\beta\rangle$ and $|\psi_1\rangle \leftrightarrow |\psi_2\rangle = |\alpha\alpha\rangle \leftrightarrow |\alpha\beta\rangle$;

Eigenvalues	Eigenfunctions	$-J \to 0$	$-J \to \infty$	
$\omega - J$	$\vert\psi_1\rangle = \vert\alpha\alpha\rangle$	$\vert\alpha\alpha\rangle$	$\vert\alpha\alpha\rangle$	(8a)
ε	$\vert\psi_2\rangle = \cos(\theta/2)\vert\alpha\beta\rangle + \sin(\theta/2)\vert\beta\alpha\rangle$	$\vert\alpha\beta\rangle$	$\frac{1}{\sqrt{2}}(\vert\alpha\beta\rangle + \vert\beta\alpha\rangle)$	(8b)
$-\varepsilon$	$\vert\psi_3\rangle = -\sin(\theta/2)\vert\alpha\beta\rangle + \cos(\theta/2)\vert\beta\alpha\rangle$	$\vert\beta\alpha\rangle$	$\frac{1}{\sqrt{2}}(-\vert\alpha\beta\rangle + \vert\beta\alpha\rangle)$	(8c)
$-\omega - J$	$\vert\psi_4\rangle = \vert\beta\beta\rangle$	$\vert\beta\beta\rangle$	$\vert\beta\beta\rangle$	(8d)
where $\omega = \frac{1}{2}(\omega_a + \omega_b)$; $q = \frac{1}{2}(\omega_a - \omega_b)$; $\varepsilon = \sqrt{q^2 + J^2}$; $\cos(\theta) = \frac{q}{\varepsilon}$; and $\sin(\theta) = -\frac{J}{\varepsilon}$				(9)

Table 1. Eigenfunctions and eigenvalues of the spin Hamiltonian Eqs. (4)–(6).

$|\psi_4\rangle \leftrightarrow |\psi_3\rangle = |\beta\beta\rangle \leftrightarrow |\beta\alpha\rangle$ convert into four distinguished EPR transitions shown in the bottom of **Figure 6**.

In **Figure 6**, the EPR transitions $|\psi_1\rangle \leftrightarrow |\psi_2\rangle$, $|\psi_2\rangle \leftrightarrow |\psi_4\rangle$ are labeled as T-type transitions because the pair spin state $|\psi_2\rangle$ correlates with the triplet $\frac{1}{\sqrt{2}}(|\alpha\beta\rangle + |\beta\alpha\rangle) = |T_0\rangle$ spin state when $-J \to \infty$. Correspondingly, the transitions $|\psi_1\rangle \leftrightarrow |\psi_3\rangle$ and $|\psi_3\rangle \leftrightarrow |\psi_4\rangle$ are labeled as S-type transitions. When $|J| \ll |q|$ the spin transitions $|\psi_1\rangle \leftrightarrow |\psi_3\rangle$ and $|\psi_2\rangle \leftrightarrow |\psi_4\rangle$ represent resonant flips of spin **a;** therefore, these transitions are labeled as the a-type transitions T_a and S_a. Similarly, $|\psi_1\rangle \leftrightarrow |\psi_2\rangle$ and $|\psi_3\rangle \leftrightarrow |\psi_4\rangle$ spin transitions are called b-type transitions T_b and S_b. The labels T_a, T_b, S_a, and S_b will be used throughout this work. In the case of the triplet precursor and absence of the TM polarization, all the three spin states $|T_0;\chi\rangle$ and $|T_\pm;\chi\rangle$ are populated equally: $\rho_{T_+T_+} = \rho_{T_0T_0} = \rho_{T_-T_-} = 1/3$. Under conditions of the non-adiabatic creation of RP, the populations of the spin states of separated pairs will be different: $\rho_{\alpha\alpha,\alpha\alpha} = \rho_{\beta\beta,\beta\beta} = 1/3$ $\rho_{\alpha\beta,\alpha\beta} = \rho_{\beta\alpha,\beta\alpha} = 1/6$. ST_0RPM CIDEP results in overpopulating of the $|\alpha\beta;\chi\rangle$ spin state in comparison to the $|\beta\alpha;\chi\rangle$ spin state for parameters used in **Figure 6**: $\rho_{\alpha\beta,\alpha\beta} = 1/6 + \delta$ and $\rho_{\beta\alpha,\beta\alpha} = 1/6 - \delta$ where $\boldsymbol{\delta} > 0$. The most remarkable thing that follows from **Figure 6** is an admixture of adiabaticity to the nonadiabatic populating of the RP spin states which lead to the same result.

3.2. History. The Closs-Forbes-Norris (CFN) model

In 1987, Closs et al. [73] suggested a model (CFN) for an SCRP diffusing within heterogeneous inhomogeneities, with sizes on the nanometer scale, for example, for micellized SCRPs or for organic biradicals with paramagnetic centers connected via a flexible tether. A highly significant distinction of the CFN model from the QSA one is that the electron spin-spin exchange interaction is considered to be a time- and space-independent "effective value," called J_{eff}. The diagram in **Figure 6** illustrating the origin of APS in terms of QSA is replaced by the spectral characteristics (**Table 1** where J is considered as time and distance independent J_{eff}) of the effective spin Hamiltonian to explain the origin of APS in terms of CFN. In fact, the CFN model operates with the averaged Hamiltonian, that is, in the limit of fast motion. This assumption greatly simplifies the interpretation of TREPR spectroscopic data collected for a confined SCRP. In calculating spectra using the CFN model, there is no need to average the spectra over the distribution of diffusional distances between radicals at the moment of observation. The model eliminates the

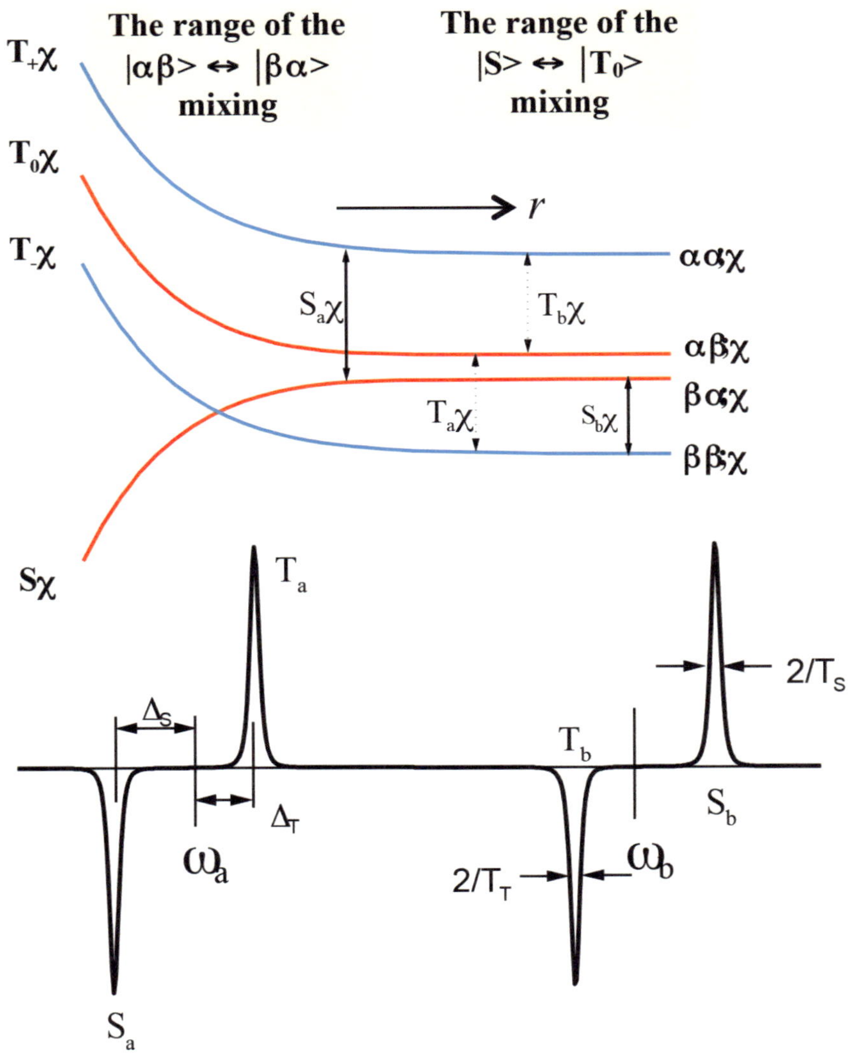

Figure 6. A correlation diagram for the spin energies of an SCRP as a function of the distance r between the radicals at a given configuration χ of the nuclear spins. The exchange potential is assumed to depend on the distance r in an exponential manner. The spectra below the correlation diagram illustrate the classification and the labeling of the TREPR transitions in a radical pair (from low to high fields, these are S_a, T_a, T_b, and S_b).

most problematic aspects of the QSA approximation. But there is a new problem—what is the effective exchange interaction J_{eff}? How is it defined? Moreover, the problems of the simultaneous occurrence of CIDEP due to the RPM, as well as the problem of line widths of the APS components, remained unsolved.

The undoubted success of the CFN model is that it is able to explain the dominance of APS in the CIDEP patterns in the TREPR spectra of confined SCRPs. Due to its physical clarity, the CFN model has enjoyed impressively wide applications [75–80].

http://dx.doi.org/10.5772/67463

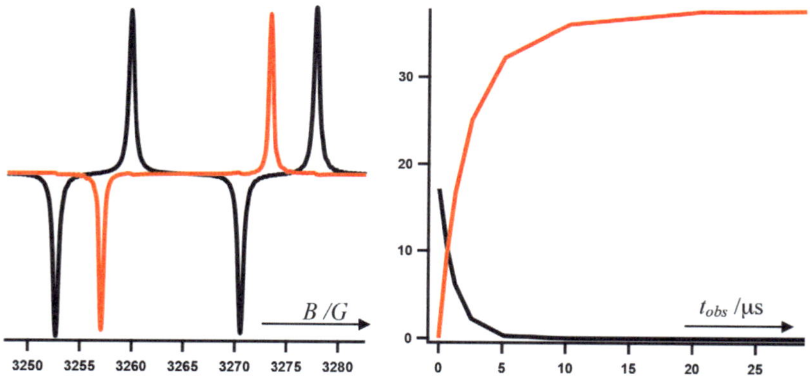

Figure 7. Decay ($k_t = 6.4 \times 10^5$ s^{-1}) of the CFN pair (solid line) and generation of the RPM–like CIDEP (dashed line). Spectral shapes of observed TREPR signals (left) at $t_{obs} = 0.8$ μs (solid) and $t_{obs} = 256$ μs (dashed); and their kinetics (right): APS- (solid) and RPM-like (dashed) CIDEP in escaped radicals. $|$Jeff/q$|$ $= 0.05$. The initial spin state of RP is assumed to be a triplet with equal populations of the $|T_+>$, $|T_->$ and $|T_o>$ spin states.

However, as new experimental observations become available, it was gradually recognized that the CFN model fit experimental data only under certain conditions. For example, rather surprisingly, J_{eff} was found to depend on temperature [81, 82] and on the chemical structure [83] of the detergent making up the micelle. The model failed to explain the dependence of the width of the spectral lines on micelle's size [82, 84, 85] or on the length of the hydrocarbon tether [86] connecting two paramagnetic centers in biradicals. Moreover, it was not clear if the importance of the compartment size arises from the space dependence of the exchange and dipolar interactions or on other factors. It must be underscored that the line widths of radical pairs were found [82, 84, 85] to be noticeably larger than those of the "escaped" ones, despite both radicals, paired and escaped, being located in the same micellar phase. This difference in line widths was found to be dependent on micellar size as well.

3.3. The CFN model: decay of the pair spin system

The assumption that the CFN spin system can decay or transform due to chemical or physical processes makes the application of the model wider and the model itself more realistic.

The initial CFN system, comprised of spins ***a*** and ***b*** and characterized by inter-radical interaction $\hat{V}_{ab}$, can be converted by a chemical reaction into a secondary CFN with a rate constant k_t. The secondary CFN is comprised of the spins ***A*** and ***B*** with the inter-radical interaction $\hat{v}_{AB}$. Such a process is described by the Liouville equation (10):

$$\begin{pmatrix} \dot{\boldsymbol{\rho}}_{ab} \\ \dot{\boldsymbol{\rho}}_{AB} \end{pmatrix} = \begin{pmatrix} -i\hat{\hat{H}}_{ab} + \hat{\hat{H}}_{\mu;ab} + \hat{\hat{R}}_{ab} - k_t \cdot \hat{\hat{E}} & \hat{\hat{0}} \\ k_t \cdot \hat{\hat{E}} & -i\hat{\hat{H}}_{AB} + \hat{\hat{H}}_{\mu;AB} + \hat{\hat{R}}_{AB} \end{pmatrix} \cdot \begin{pmatrix} \boldsymbol{\rho}_{ab} \\ \boldsymbol{\rho}_{AB} \end{pmatrix} \quad (10)$$

If radicals ***A*** and ***B*** do not interact ($\hat{v}_{AB} = 0$), then they become free radicals or "escaped." From Eq. (10), the corresponding TREPR spectra are shown in **Figure 7**. The TREPR spectrum of the new CFN pair (**Figure 7**, left-hand, dashed line), comprised of non-interacting radicals, is identical to that what is called CIDEP due to the ST_0RPM. In fact, in this calculation, we have reproduced the so-called exponential model of CIDEP. This model is considered to be absolutely unrealistic for the generation of CIDEP in a chemical reaction resulting in geminate radical pairs, but here, we deal with a CFN pair as a precursor and its conversion to the secondary one. For the CFN pair, the effective exchange interaction is allowed to be arbitrary small, which makes it probable that the decay of the CFN polarization is followed by the creation of CIDEP, similar to what is induced by the exponential model like CKO as applied to CIDEP.

4. Diffusion models for APS

4.1. SCRPM versus ST_0RPM

In order to address some of the deficiencies inherent to the QSA, in 1991, Shushin [38] suggested a diffusive theory for SCRP-induced CIDEP in homogeneous solvents. Instead of the QSA, Shushin considered a solution to the Stochastic Liouville Equation (SLE) as applied to freely diffusing radicals. The evolution of the SCRP total spin ($\hat{\mathbf{S}} = \hat{\mathbf{S}}_a + \hat{\mathbf{S}}_b$) is governed by the spin Hamiltonian Eqs. (4)–(6). The main difference between the QSA and SLE approaches is that the averaging of the spectra corresponding to the instantaneous spin Hamiltonians in the QSA is replaced by an averaging of the pair-spin density matrix over stochastic diffusional processes governed by the SLE.

The model is correct for $t_{obs} \gg T_2$ or for $\bar{r} \gg \sqrt{D T_2}$, where $\bar{r}$ is average distance between radicals at $t = t_{obs}$ and $D = D_a + D_b$ is the coefficient of their mutual diffusion. Shushin's model is much more than a refined treatment of the problem. In fact, it drastically changed the qualitative interpretation of the APS by experimentalists. Now, not only do the interacting radicals contribute to the APS but also there are additional contributions from radicals which interacted in the past and yet do not interact at the moment of observation. This is seen from the analytical expression [38] for the line shape of the TREPR signal at t_{obs}:

$$\begin{aligned} I(\omega) \cong\ & 2\pi \cdot P_{ST}\left(\frac{T_2}{1+T_2^2(\omega-\omega_a)^2} - \frac{T_2}{1+T_2^2(\omega-\omega_b)^2}\right) \\ & - 4\pi^2 P_0 D\lambda \cdot \left[\frac{T_2^3(\omega-\omega_a)}{\left(1+T_2^2(\omega-\omega_a)^2\right)^2} + \frac{T_2^3(\omega-\omega_b)}{\left(1+T_2^2(\omega-\omega_b)^2\right)^2}\right] sign(J) \end{aligned} \tag{11}$$

where P_{ST} is the enhancement factor of CIDEP due to ST_0RPM and P_0 can be interpreted as a concentration of contact SCRPs. The term $1/T_2$ is the Lorentzian line width, including exchange broadening. From Eq. (11), the TREPR spectrum of a SCRP is comprised of two distinguishable parts: (1) multiplet ST_0RPM polarized Lorentzian EPR signals centered at the resonance magnetic fields of the individual free radicals (ω_a and ω_b) and (2) two APS-split signals which are simply the first derivatives of the Lorentzian lines centered at the individual free radical resonances. The amplitude of these APS-split signals is proportional to the concentration of SCRPs present at t_{obs} that still have an opportunity for a re-encounter. The radicals that have lost this opportunity contribute to the ST_0RPM polarization.

What immediately follows from Shushin's diffusion model is the dominance of the polarization due to ST_0RPM, even at the shortest possible times of observations (~100 ns). Using representative parameters, the relative intensity of the APS is no more than 0.1%. Such a small distortion of the ST_0RPM polarized signal cannot be discerned in an ordinary *cw* TREPR experiment. It can therefore be stated with confidence that except for the central line of a spectrum, where the ST_0RPM contribution approaches zero, it is impossible to detect an APS in ordinary liquid solutions.

At first glance, this conclusion contradicts many experimental observations [87]. However, Tominaga et al. [87] came to the conclusion that at low temperatures (< −70°C), there exist solvent/solute structures in *i*-propanol/acetone mixtures which persist for a few microseconds and that the guest ketyl radicals are fixed in these structures by means of hydrogen bonding to the host *i*-propanol molecules. This is in reasonable agreement with the suggestion [88] that geminate radical pairs can be trapped in a spherically symmetric potential hole and diffuse freely within it.

4.2. The Microreactor model for the micellized radical pair

By itself, the microreactor model [89, 90] was formulated to explain [91] the extremely high efficiency [92] of the magnetic isotope separation in chemical reactions of geminate radical pairs confined in a micellar phase. Essentially, the model is a synthesis of mathematical formalism [31–33] developed to account for the spin dynamics of radical pairs, with a particular model for diffusion-controlled reactions in micellar media [93, 94]. To make the model workable, we make two assumptions: (1) the microreactor model approximates the micelle as a spherical homogeneous drop of radius L_m (**Scheme 4**) and (**2**) One radical from the pair, of radius r_a, is considered to be fixed at the center of the micelle, while the other, of radius r_b, is allowed to diffuse and to escape from the micelle to the water bulk.

The theory of the TREPR spectroscopy of SCRP is described in details in references [84, 85].

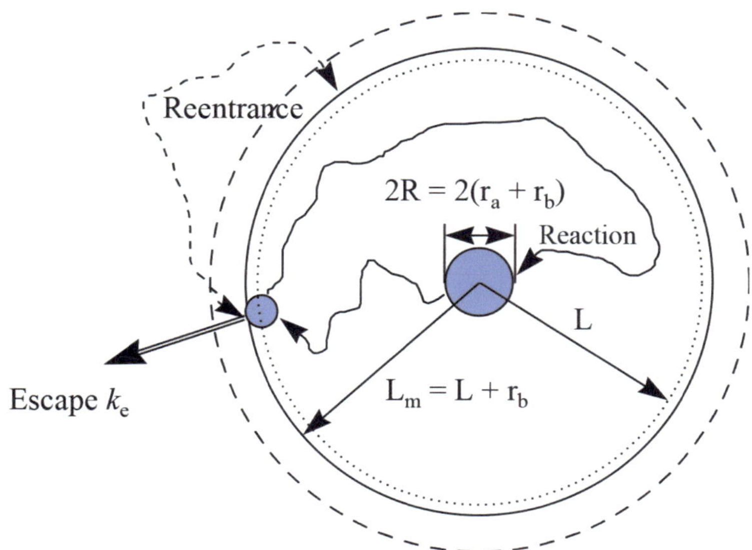

Scheme 4. Visualization of the microreactor model for SCRPs.

5. Micelle size dependence

5.1. Photolysis of methyl desoxybenzoin (MDB) [84]

Here, we present an example that demonstrates the high sensitivity observed for the APS in the ^{13}C-benzoyl radical as a function of the size of its alkyl sulfate micellar host. Moreover, it has been found that a decrease in micelle size results in a strong asymmetry of the observed APS.

Upon photoexcitation (**Scheme 5**), the ketone MDB dissociates into a triplet SCRP consisting of benzoyl and sec-phenethyl radicals (**Scheme 5**).

The TREPR spectra obtained for this system at $t_{obs} = 500$ ns after the laser flash ($\lambda = 308$ nm) are shown in **Figure 8**. The dominant feature here is the $m_N = \pm½$ doublet from the ^{13}C-benzoyl radical, with a hyperfine coupling constant on the ^{13}C carbonyl, A(^{13}C) = 127 G. Each line in this doublet is a superposition of the APS from the SCRP and ST_0RPM polarized signal from free (escaped) radicals. In the latter, the $m_N = +1/2$ line is in emission; the corresponding $m_N = -½$ line is in absorption (the E/A pattern). The SCRPM contribution dominates at the early delay times and in large micelles. The ST_0RPM is more prominent at the longer delay times and in small micelles. Note that as predicted above, the broad APS-split signals show no ST_0 polarization.

The Δ_{APS} splitting (defined as the distance between the extremes of the APS) was found to increase when the micelle size decreases: 0.85 mT in SDS (C_{12}), 1.42 mT in undecyl sulfate (C_{11}), and 1.58 mT in decyl sulfate (C_{10}) micelles. Also, the two lines comprising the APS have different intensities. Except for the C_{12} micelles, where the APS is almost symmetric, the T-type lines are stronger and narrower than the S-type partners. This asymmetry is more prominent in smaller micelles. The shape of the APS pattern was found to be independent of the delay time t_{obs} and the amplitude of microwave field ($\omega_1 = (0.01 - 2) \times 10^6$ rad/s). The E/A-polarized lines of the benzoyl radicals are much narrower than the width of the APS components (0.06 mT vs. 0.8–1.5 mT).

The parameters used in our calculations using the microreactor model are given in **Table 2**, where σ is the boundary factor describing the probability for radicals to escape from the micelle core and P is the reaction probability of the SCRP ($P = P_r + P_d$; to elaborate, the terms P_r and P_d are the probabilities for recombination and disproportionation of the radicals, respectively). We can measure these values directly using racemization of the optically active MDB and by measuring the chemical yield of benzaldehyde. The term $Z = \frac{3RD}{L^3 - R^3}$ [84] is the

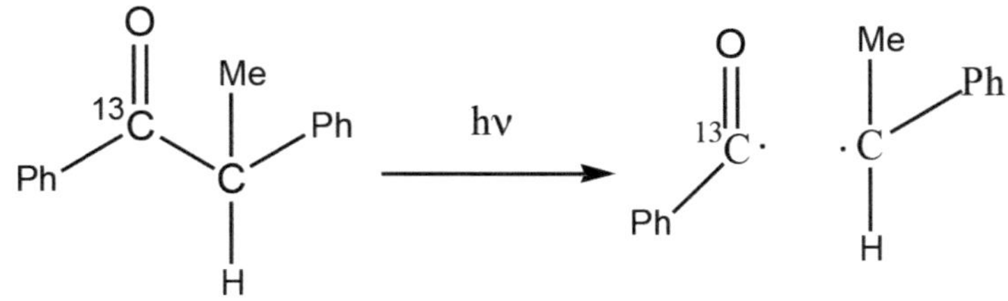

Scheme 5. Photolysis of MDB results in formation of the triplet SCRP consisting of benzoyl and sec-phenethyl radicals.

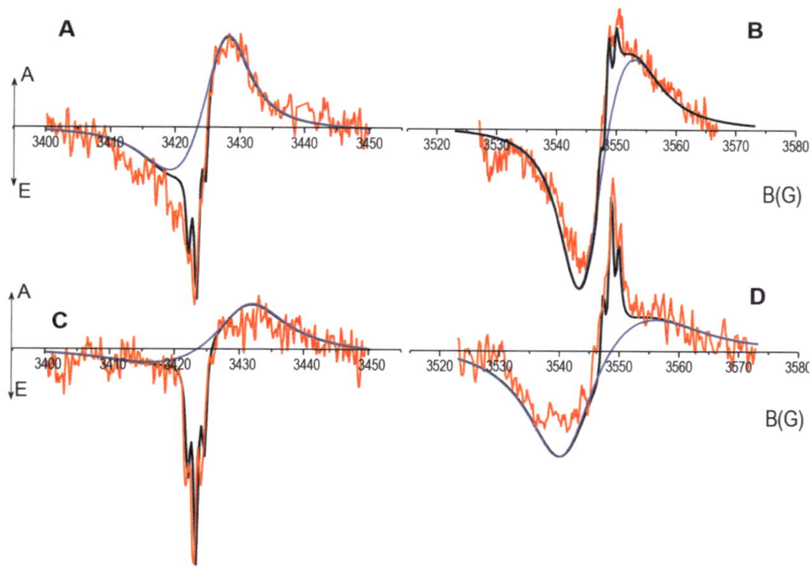

Figure 8. TREPR spectra observed in laser photolysis of MDB in sodium dodecyl (A and B) and in sodium decyl (C and D) sulfate micelles (open circles). Only $M_N = +½$ (A and C) and $M_N = -½$ (B and D) signals from ^{13}C–benzoyl radicals are shown (hfi constant A(^{13}C) = 127 G). The transverse magnetization *M(t)* was convoluted with $f_1(t) = (t/\tau_w^2)exp(-t/\tau_w)$; $\tau_w = 500$ ns according to window fuction used in simulations. Simulated spectra (bold lines) were calculated using the model of microractor. Dashed lines represent TREPR spectra of radical pairs inside the micelles.

n^a	Δ_{APS} (mT)	$L \times 10^8$ cm^b	$D \times 10^6$ cm^2/s	σ	P^c	$Z \times 10^{-7}$ s^{-1}	$J_o\tau_{EX}$ ^{d}	$J_0\lambda^2/D$	$A/4Z$
12	0.85	15.4	1.58	0.022	0.51	8.28	2.43	0.20	26.6
11	1.42	14.2	2.02	0.018	0.53	13.7	1.90	0.16	16.1
10	1.58	12.9	2.50	0.015	0.51	23.3	1.54	0.13	9.44

[a]The number of carbon atoms in the detergent chain.
[b]The choice of the value of L is discussed in [94]; note, that the actual micellar size $L_m = L + r_b$ (see **Scheme 4**).
[c] This is total reaction probability including the recombination and disproportionation.
[d]$J_0 = -2.2 \times 10^{10}$ rad/s, $\lambda = 0.5 \times 10^{-8}$ cm, $R = 6 \times 10^{-8}$ cm.

Table 2. Parameters used in simulations in **Figure 8** and representative values.

frequency of forced encounters of radicals in the micellar phase; $J_0\tau_{ex}$ [84] and $J_0\lambda^2/D = J_0\tau_c$ [84] are parameters characterizing the efficiency of spin exchange.

5.2. Photolysis of dimethyl desoxybenzoin (DMDB) [81]

The photolysis of DMDB is illustrated in **Scheme 2**. The TREPR spectra acquired at $t_{obs} = 300$ ns are shown in **Figure 9**. Qualitatively, the observed spectra from MDB and DMDB are similar (compare **Figures 8** and **9**). However, the quantitative conclusions are different. Micelle sizes extracted from simulations of SCRP and TREPR spectra, resulting from the photolysis of MDB and DMDB are practically same as that for SDS micelles—$L = 15.4$ Å for MDB and 15.7 Å for

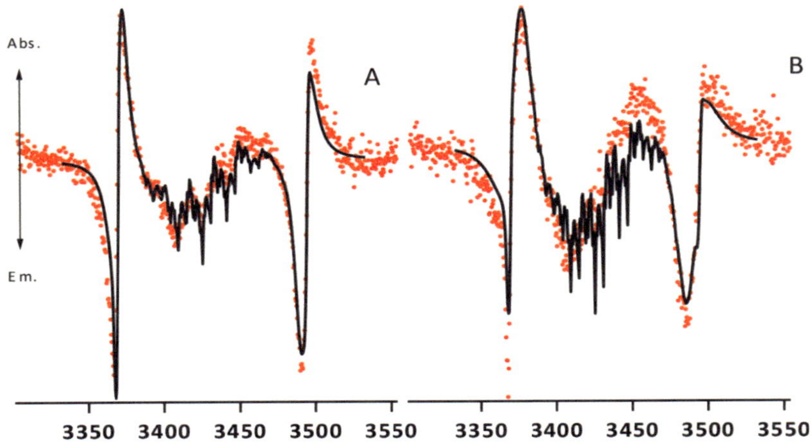

Figure 9. Time–resolved ESR spectra observed (open circles) in 308 nm laser photolysis of DMDB (2 mM): (A) in SDS (0.1 M); (B) in decyl sulfate (SDeS) (0.21M) micelles. The calculated spectra are represented by solid lines. Parameters used for the simulations are; $J_0 = -2.4 \times 10^{10}$ rad/s, $\lambda = 0.5$ Å, $R = 6$ Å, L(SDS) = 15.7 Å, L(SDeS) = 11.6 Å, D(SDS) = 0.7×10^{-6} cm^2/s, D(SDeS) = 1.2×10^{-6} cm^2/s. (a – ^{13}C–benzoyl) A_a[CO] = 127 G, $g_a = 2.0006$, $g_b = 2.0026$; $T_{1a} = 0.1$ μs, $T_{1b} = 3.0$ μs, $T_{2a} = 0.08$ μs, $T_{2b} = 1.6$ μs, $\tau_w = 0.3$ μs.

DMDB. However, in the case of SDeS, they differ significantly: $L = 12.9$ Å for MDB and 11.6 Å for DMDB. Diffusion coefficients were found to be different as well: $D = 1.08 \times 10^{-6}$ cm^2/s and 2.5×10^{-6} cm^2/s in SDS and SDeS in the case of MDB and respectively, 0.7×10^{-6} cm^2/s and 1.2×10^{-6} cm^2/s for the case of DMDB.

Comparing these values, one has to keep in mind that the sec-phenethyl radical is more hydrophilic and less bulky than the cumyl radical. This easily explains the differences obtained for the diffusion coefficients. As far as the micelle size difference is concerned, the difference in hydrophobicity is the main reason. Indeed, the higher the hydrophobicity of the radical, the smaller the range available for radical diffusion inside the micelle. Of course, the smaller the micelle size, the stronger the influence of this factor.

At the present stage of both experimental and computational analysis, it is rather difficult to make conclusions as to how we could estimate the importance of these differences between MDB and DMDB. Further investigations are necessary. Nevertheless, the observed differences give a fairly good reason to consider the TREPR spectroscopy of spin-correlated radical pairs to be a very sensitive method for studying diffusional mobility and the characteristic sizes of the inhomogeneities where the spin-correlated pairs are localized.

5.3. Photolysis of (2,4,6-trimethylbenzoyl)-diphenylphosphine oxide (TMDPO) [85]

The photoexcitation (**Scheme 6**) of TMDPO leads to the formation of a geminate triplet SCRP of diphenyl phosphonyl and 2,4,6-trimethylbenzoyl radicals via dissociation of a short-lived triplet state (lifetime less than 1 ns) of TMDPO.

There are three distinctive features of the spectra in **Figure 10**: (1) the diphenyl phosphonyl radical possesses one of the largest known hyperfine coupling constants (385 G) in the family of "organic" radicals. This allows for the manifestation of the so-called ST_RPM CIDEP [85],

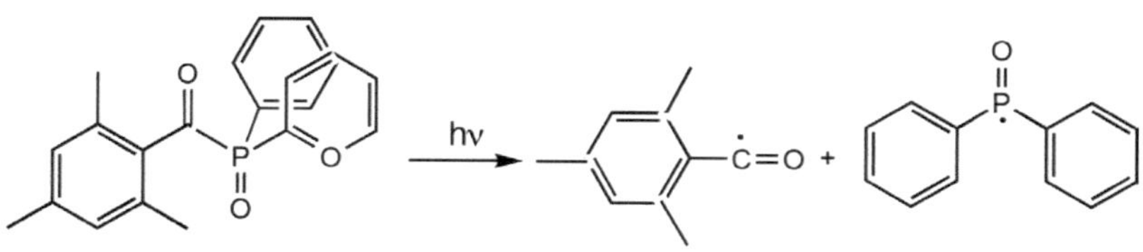

Scheme 6. Photolysis of TMPBPO.

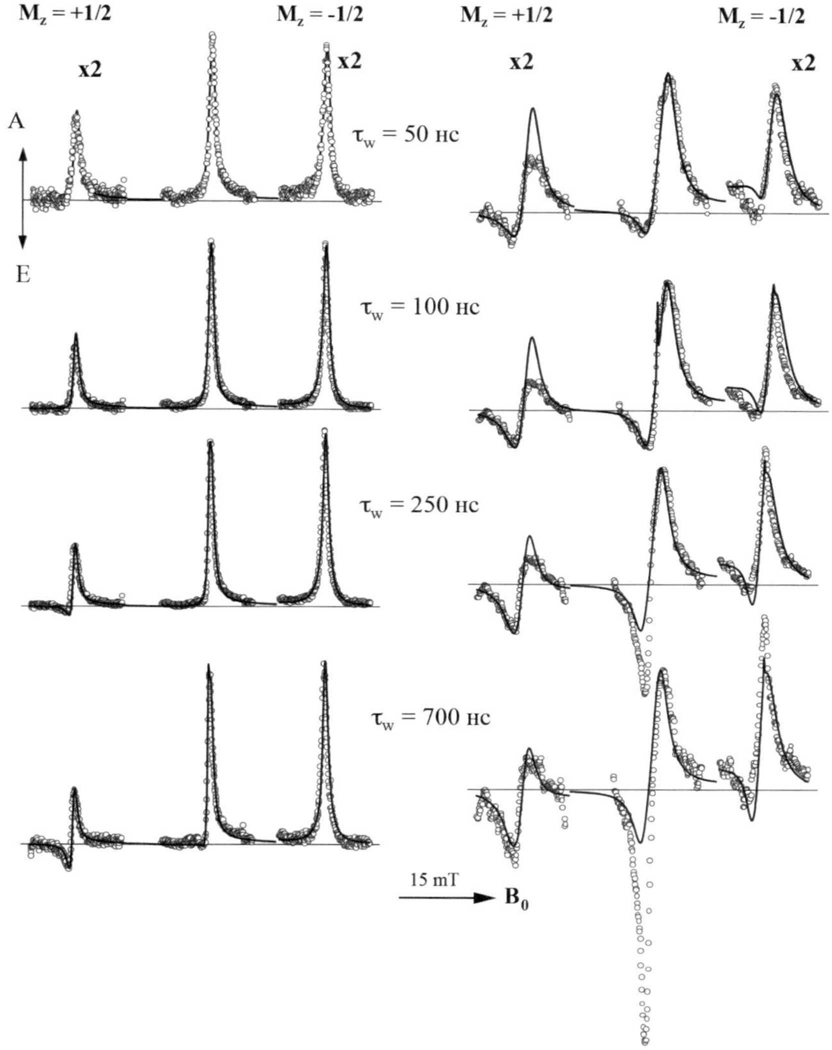

Figure 10. The TREPR spectra observed during 308 nm laser photolysis of TMDPO in SDS (C_{12}) and in sodium octylsulfate (C_8) micelles. The parameter t_w is the gate width of the t-exponential window. The intensities of the $m_P = \pm 1/2$ lines from the diphenyl phosphonyl radical have been enlarged by a factor of two for convenience. The simulated spectra are calculated using the microreactor model and are shown by solid lines. Flip-flop electron-nuclear spin transitions were taken into account to simulate ST_RPM polarization with a distance-dependent exchange potential. Parameters used in this simulation are given in **Table 3**. The term Δ_{APS} is defined as the splitting, measured in Gauss, between the extremes of the APS.

even in the relatively high-field X-band experiment (0.34 T). (2) This is the first observation of a competition between the SCRPM and the ST_RPM. We have already mentioned that the ST_0RPM CIDEP is not seen when the SCRPM dominates; SCRPs from MDB and from DMDB photolysis do not show a ST_0RPM contribution (**Figures 8** and **9**). But at the same time, escaped radicals demonstrate ST_0RPM CIDEP and do not show any SCRPM (APS). It is rather intriguing that there is competition between the SCRPM and the ST_RPM and yet simultaneously no competition between the ST_0RPM and the SCRPM. (3) At initial observation times, the pair is strongly TM polarized. The evolution of the TM CIDEP to the ST_RPM through the SCRPM is extremely sensitive to the micelle size, as **Figure 10** convincingly demonstrates.

It is also interesting to note that changing parameters such as diffusion coefficients and microreactor sizes follows the same logic established by comparing these parameters to the cases of MDB and DMDB (see **Figure 9**, **Table 2**, and surrounding discussion).

5.4. Photooxydation of glycyl-glycine (GG) by the electronically excited triplet state of anthraquinone-2,6-disulfonate (AQDS) in AOT reverse micelles [95]

Figure 11 shows the TREPR spectra obtained during the photoexcitation of anthraquinone-2,6-disulfonate (AQDS) in the presence of diglycine (GG) where both the photosensitizer and the substrate are confined to the aqueous interior of AOT reverse micelles (**Scheme 7**). A remarkable feature of these spectra is the observation of the superposition of two CIDEP patterns, APS and ST_0RPM. The ST_0RPM polarization observed here does not have the appearance of spectral lines expected of free radicals as an addition to APS (cf. **Figure 8**). In this particular case, the contribution of ST_0RPM is seen as an increase in the intensities of the S-type components of APS, that is, the ST_0RPM polarization belongs to the SCRP indeed but not to the free radicals that have escaped from the water core into the bulk solvent (n-octane), with subsequent localization in other reverse micelles.

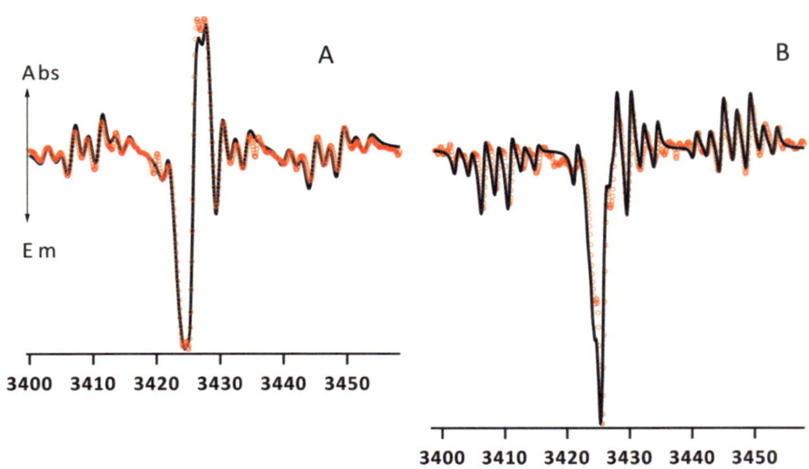

Figure 11. TREPR spectra (empty circles) observed under photooxidation (laser flash photolysis, $\lambda = 308$ nm) of GG by AQDS in AOT micelles for L_m values of 23 Å (A) and 53 Å (B), at room temperature. Solid lines are simulations in terms of the microreactor model.

http://dx.doi.org/10.5772/67463

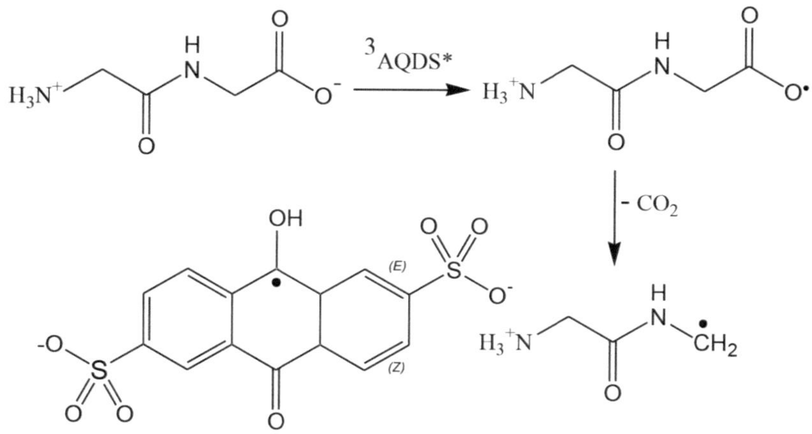

1 **2**

Scheme 7. Generation of geminate triplet radical pair 1…2 under the photolysis of AQDS in AOT reverse micelles.

Simulations of the TREPR spectra of SCRPs in alkyl sulfate micelles showed that the viscosity of the micellar phase decreases with the micelle size. AOT micelles seem to behave differently. A decrease in the water core radius from 53 Å to 17 Å causes a decrease in the mutual diffusion coefficient from 2.4×10^6 cm^2/s to 0.05×10^6 cm^2/s. This is in reasonable agreement with other measurements reported from fluorescence depolarization experiments [98], that is, the aqueous interior of reverse micelles becomes more viscous as its size decreases. It is speculated that the decrease in size is accompanied by a greater degree of ordering of the solvent.

Another remarkable observation is that the ST_0RPM contribution dominates in the micelles with an extremely low frequency of forced encounters $Z = 5.2 \times 10^6$ s^{-1}, compare with the lowest frequencies $Z = 8.28 \times 10^7$ s^{-1} in the case of MDB in SDS and $Z = 2 \times 10^7$ s^{-1} in the case of TMDPO in SDS. Whether this finding makes physical sense is less important, at least at present, than recognizing that the peculiar distortion of the APS by the ST_0RPM can be a very sensitive measurement tool for the sizes of inhomogeneities on the nanometer scale.

6. Temperature dependence: photoreduction of benzophenone in SDS micelles [82, 96]

From the point of view of the CFN model and QSA, the temperature dependence of the APS is quite unexpected. Indeed, the QSA operates in terms of the distance dependent on exchange potential and always in regard to a particular distribution of radicals at the time of observation. Neither of these can change the shape of APS but both can drastically alter its intensity. The CFN model utilizes the concept of an effective exchange interaction, but the model's temperature dependence is not immediately clear.

The radical structures under consideration are shown at the top of **Scheme 8**. The photoreduction of the excited triplet state of BP (^{3}BP*) in SDS [97] results in a triplet-born SCRP consisting of

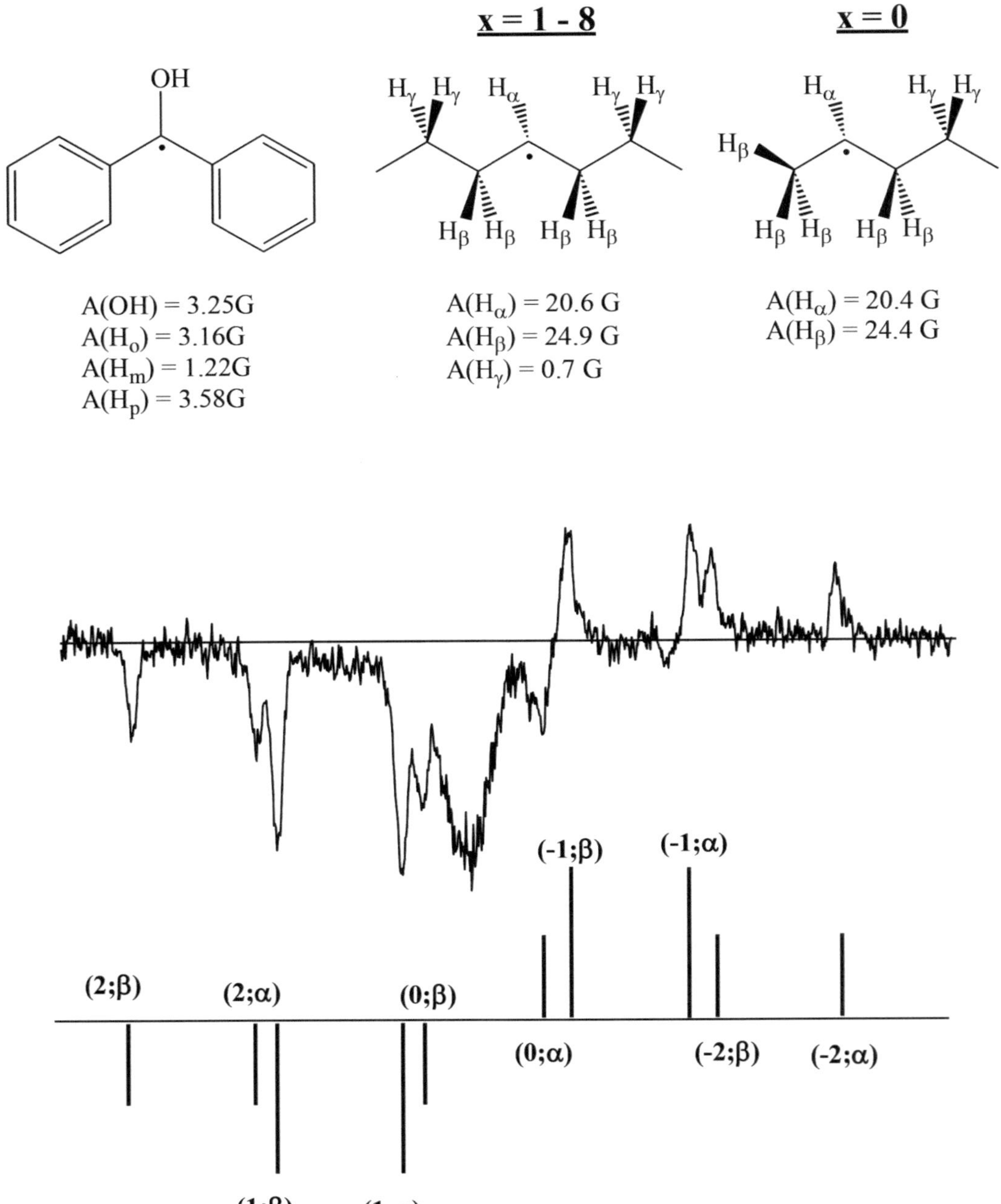

Scheme 8. Stick plot (bottom) of the TREPR transitions expected for the primary alkyl radical products from the photochemical reduction of the alkyl chain of a surfactant by a triplet sensitizer (top). A TREPR spectrum of the escape radicals of SDS is shown in the middle.

http://dx.doi.org/10.5772/67463

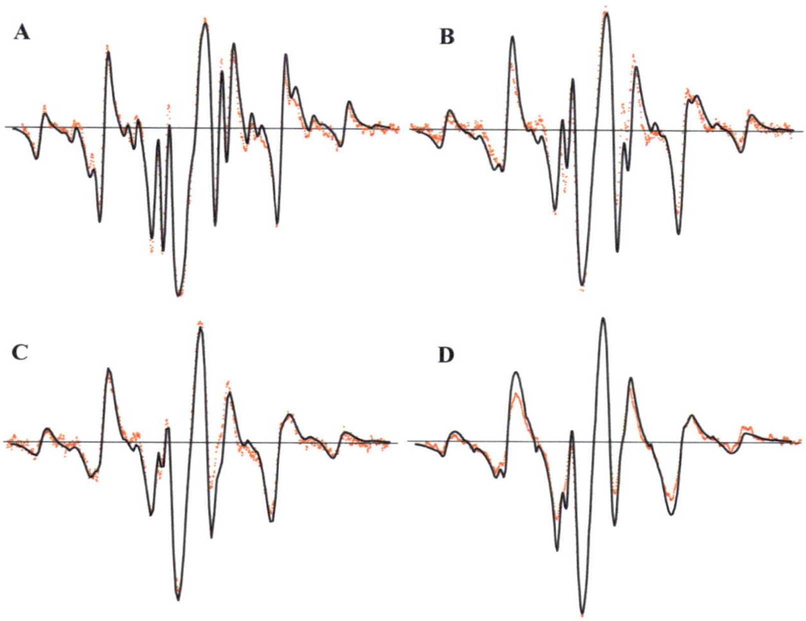

Figure 12. TREPR spectra observed during 308 nm laser photolysis of BP/SDS (open circles), acquired at a 500 ns delay time (gate width of the boxcar window was 100 ns), at 16°C (A), 32°C (B), 46°C (C), and 60°C (D). Simulated spectra (the t–exponential sampling window is centered at 500 ns) were calculated using the microreactor model and are designated by solid black lines. Parameters used in the simulations are listed in **Table 5**.

surfactant-based secondary alkyl radicals of the general structure $CH_3(CH_2)_n\dot{C}H(CH_2)_{10-n}OSO_3^-$ ($n = 0$–9) and the BP ketyl radical ($Ph_2\dot{C}OH$). **Figure 12** shows simulations of the experimental spectra for this system in terms of a numerical solution of the SLE as applied to the microreactor model. Despite some rather simple but necessary simplifications, the model is capable of reproducing all of the key features of the experimental spectra and all of the observed temperature effects. Specifically, these are [82, 91]: (1) increasing the temperature causes broadening of the APS components, (2) the strong line shape asymmetry in the APS spectral pattern depends on the q values, (3) narrowing of the central portion of the spectra takes place with increased temperature, and (4) narrowing of the line widths of the escaped alkyl radicals is observed. The parameters used in the calculations are presented in **Table 3**. All the values are in reasonable agreement with available data from other experiments. In general, the temperature effect was very similar to that of reducing the size of micelles. This suggests that molecular translational mobility, and its restrictions, is the key factor in determining the spectral shape of the APS in the TREPR spectra of SCRPs experiencing limited mobility. The dependence of the asymmetry phenomenon on the value of q is a fundamental issue to be addressed for future applications of SCRPs as spin-polarized probes.

We have successfully demonstrated that the TREPR spectra of SCRPs, coupled with simulations in terms of the microreactor model, can be used to investigate the sizes and internal viscosities of

Temp. $\pm 0.5°C$	L^b (Å)	$D \times 10^{-7}$ (cm^2/s)	$A[H_\beta]^c$ (G)	$T_{2a}{}^d$ (μs)	$k_{rel} \times 10^8$ (s^{-1})^e	$k_g \times 10^6$ (s^{-1})^e	$Z \times 10^7$ (s^{-1})	$J_0\lambda R/D$ ($J_0\lambda^2/D$)	$q(0,\alpha)/Z$
16	17.2	2.0	24.9	0.080	2.0	2.2	0.74	36 (3)	3.1
32	16.3	3.5	24.6	0.095	3.3	4.5	1.5	21 (1.8)	1.5
46	15.6	5.3	24.4	0.107	4.8	7.8	2.7	14 (1.2)	0.84
60	14.6	8.0	24.2	0.120	6.8	13	4.8	9.0 (0.8)	0.47

[a]Temperature independent parameters: $J_0 = -2.4 \times 10^{10}$ rad/s; $\lambda = 0.5$ Å; $R = 6$ Å; $D_{ZFS} = 3.4 \times 10^{10}$ rad/s [41]; $\sigma = 0.024$; $\Lambda_s = 0.62$; $\eta = 0.2$; $g_a = 2.0026$; $g_b = 2.0028$; $\omega/2\pi = 9.8$ GHz; and $\omega_1 = 10^5$ rad/s.
[b]The variation of L within 0.4 Å at 16°C and 0.2 Å at 60°C is fairly acceptable. In **Table 1**, the upper limit values are presented. All other parameters are those that have been used with the presented L values. Note that L is the radius of spherical volume available for radicals. Thus, the real micelle size is greater by approximately 3 Å.
[c]The hyperfine coupling constant with the H_α protons in the $n = 1$–8 fragments do not seem to vary noticeably with temperature.
[d]T_{2b} was assumed to be the same (see text). T_{1a} and T_{1b} were also assumed to be the same and independent of temperature. It is likely that this is incorrect, but the spectral shape is insensitive to these parameters when they are varied within reasonable limits.
[e]k_g and k_{rel} are the rates respectively of the hydrogen abstraction and relaxation of the polarized triplet state of benzophenone.

Table 3. Parameters[a] used to simulate the TREPR spectra of alkyl/ketyl RPs in SDS micelles at different temperatures.

inhomogeneous structures on the nanometer scale. Such experiments convincingly show that they can be a source of valuable information on the inter-radical interactions as well. However, even successful simulations bring an aesthetic satisfaction without increasing our knowledge of the physical characteristics of the system. In the next section, we will try to resolve this deficiency.

7. Random walks in micelles: ST_0RPM versus SCRPM

Our experimental observations and their successful simulations in terms of the microreactor model teach us that in alkylsulfate micelles, the ST_0RPM polarization is observed only for escaped radicals, while in AOT reverse micelles, the ST_0RPM can be either negligible or dominating. In homogeneous solution, the ST_0RPM undoubtedly dominates the TREPR spectra. This is easily demonstrated using the microreactor model for large values of L_m. This point is crucial, as the early theories of CIDEP addressed precisely the case of radicals that never experienced repeated collisions after they were involved in the creation of polarization. Mathematically, this is realized through the calculations of reduced spin density matrices by the Tr operation. Calculations in terms of the microreactor model show that the observed ST_0RPM polarization is present not only in the radicals released from the medium but also on those inside of it, if the values of L_m and the viscosity of micelle core are sufficiently large (compare, for instance, the values of these parameters given in **Tables 4** and **5**).

Analytical solutions to the problem of mutual diffusion of two particles, within a closed volume on the nanometer scale, exist only for the particular case of a spherically symmetric compartment, with one of the partners fixed in the center of the sphere [89, 90, 93]. Instead, we apply the Random Walks (RW) model to integrate the SLE as applied to an SCRP confined in a

n	Δ_{APS}[a] (mT)	k_{obs}[a] $\times 10^7$ (s^{-1})	L^b (nm)	$D^b \times 10^6$ (cm^2/s)	η^b	T_{1a}[b] (ns)	$Z^c \times 10^{-7}$ (s^{-1})	$-J_0\tau_{ex}$[c]
12	0.85±0.04	1.0±0.2	16.0	0.44	0.44	90	2.70	30.0
11	1.19±0.04		15.2	0.51	0.41	80	3.70	25.9
10	1.60±0.04	1.9±0.2	14.4	0.58	0.38	70	5.03	22.8
9	2.05±0.04		13.7	0.65	0.35	60	6.81	20.3
8	2.47±0.06	2.4±0.2	13.0	0.72	0.32	50	9.13	18.3

[a]k_a is the rate constant of the SCRP decay. It was measured in the laser flash ($\lambda = 308$ nm) photolysis experiments.
[b]From the best fit of TREPR spectra. Note that in this case, the values of L are slightly larger than those used in simulations of the TREPR spectra of SCRPs generated in photolysis of MDB and DMDB. This is the most likely to be a consequence of the fact that the SCRP in the case of TMDPO is more hydrophilic.
[c]Calculated from the parameters used for simulation.
[d]Other parameters used in calculations: $J_0 = -4.4\times10^{10}$ rad/s, $\lambda = 0.04$ nm, $A = 38.5$ mT, $g_a = 2.0034$ (a – diphenylphosphonyl), $g_b = 2.0006$ (b – 2,4,6, trimethylbenzoyl); and η is the TM polarization $\eta = (\rho_{T_+T_+} - \rho_{T_-T_-})/(\rho_{T_+T_+} + \rho_{T_-T_{-+}})$; $\omega_0 = 9.766979$ GHz, $\omega_1 = 10^6$ rad/s, $T_{2a} = 40$ ns, $T_{1b} = 1$ μs, $T_{2b} = 160$ ns, $k_s\tau = 0.58$, $k_d = 10^6$ s^{-1}, and $R = 0.75$ nm.

Table 4. Observed APS splittings and Δ_{APS}, and decay constants k_{obs}, and the sets of parameters used to simulate time-resolved ESR for the pair of diphenylphosphonyl/2,4,6-trimethylbenzoyl radicals in C_n alkyl sulfate micelles[d].

$[H_2O]/[AOT]$	Water core radius (Å)	J_0 (GHz)	D ($\times 10^6$ cm^2/s)	T_2 (μs)	Z ($\times 10^{-6}$ s^{-1})	$\frac{qR^2}{D}$
37	53	−3.1	2.4	0.36	5.2	0.32
	43	−3.1	1.6	0.28	8.0	0.49
23	33	−3.1	1.2	0.24	16	0.65
16	23	−3.1	0.4	0.24	30	2.0
12	17	−3.1	0.05	0.12	27	16

[a]Other parameters: g-factors of the radicals **1** and **2** respectively: $g_1 = 2.0029$, $g_2 = 2.0041$. HFI constants in radical **1**: $A_1(CH_2) = 1.91$ mT, $A_2(CH_2) = 0.422$ mT, and $A(N) = 0.22$ mT; in radical **2**: $A_1(H) = 0.121$ mT, $A_2(H) = 0.042$ mT, and $A_3(H) = 0.036$ mT; rate constant of spin-non-selective decay of radical **1** $k_1 = 1.2 \times 10^6$ s^{-1}; the probability of the singlet contact radical pair decay 0.52 was held as independent on the micelle size, maximum of the time-window function was located at 500 ns.

Table 5. Parameters used in the microreactor model simulation and the frequencies of radical encounters extracted from simulations for **Figure 11**[a].

nanoscopic cavity, that is, to integrate Eqs. (4)–(6). In such a way, we are free from the restrictions imposed by the demands of a spherical shape for the microreactor, and we can also relax the requirement of fixing one of the radical partners in the center of microreactor.

The space evolution of a pair can be explicitly calculated (**Figure 13**) in the form of trajectories $\Omega(\mathbf{s}(t))$, where $\mathbf{s}(t)$ stands for space stochastic variables limited by the demand that particles cannot penetrate into each other and cannot escape from the cavity. Instead, they can decay through a spin-selective chemical reaction with a rate constant that depends on the distance between the radicals. Then, the propagation of the spin density operator $\hat{\rho}_\Omega(t)$ of SCRP is given by the formal solution of Eqs. (4)–(6) under condition that the interval Δt is so small that all the parameters of SLE can be considered as time independent for this time interval:

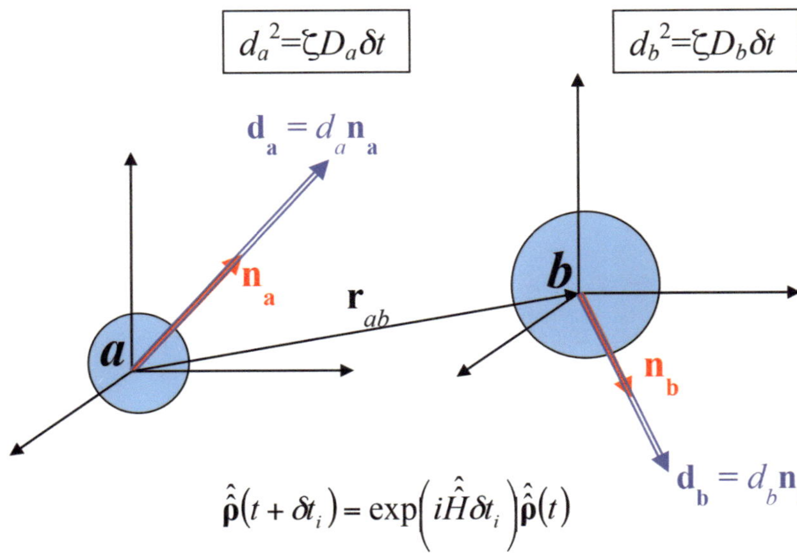

Figure 13. Random walk calculations: Each radical ***a*** and ***b*** take walks along random directions $\mathbf{n_a}$ and $\mathbf{n_b}$, ζ is the coefficient (=6 for the 3- and = 4 for the 2-dimension) depending on the dimensionality of walk. If, say, radical ***a*** attempts to penetrate into the sphere (= r_b) representing radical ***b***, then it will return to its initial position and this pair will be considered as a reacting pair according to the rules for spin-selective reaction decay. Similarly, if radical ***a*** or ***b*** attempts to cross the boundary, it will return to the starting point as well. During the time interval δt, the spin dynamics of the pair is governed by the pair spin Hamiltonian, Eq. (4), depending on the distance between the radicals r_{ab}.

$$\hat{\rho}_\Omega(t+\Delta t) = \exp\left\{\left(-i\hat{H}\left(\Omega(t)\right) - \hat{K}\left(\Omega(t)\right)\right)\Delta t\right\} \times \hat{\rho}_\Omega(t)$$
$$\times \exp\left\{\left(+i\hat{H}\left(\Omega(t)\right) - \hat{K}\left(\Omega(t)\right)\right)\Delta t\right\} \tag{12}$$

It is clear that the number of trajectories needs to be large, and the time intervals Δt should be small enough such that $|\Delta J| = |J(t+\Delta t) - J(t)| \ll |q|$. Control over the implementation of this inequality greatly accelerates the speed of our calculations [99]. We use the "crude force" method (or a capture-probability based algorithm) to investigate the steady state conditions and the kinetics of approaching it. Such a choice is dictated by that under conditions of spin–selectivity of reaction application of advanced algorithms, [100] is rather difficult. An essential drawback of the method is the practical impossibility to check convergence of the solution.

7.1. The concept of the filled-out micelle and enforced encounters

Immediately after birth at $t = 0$, the radicals of the SCRP are assumed to be at the distance of the closest approach R which can be estimated as 5–10 Å (in our calculations we always use $R = 6$ Å, except in the case of TMDPO, where $R = 7.5$ Å). Then, until one of the radicals arrives for the first time at the boundary of micelle, the radicals diffuse in such a way if they were free, that is, unrestricted by micellar containment. After this moment, it is quite reasonable to utilize, as the first approximation, the frequency Z of encounters of the SCRP partners confined in a micellar

http://dx.doi.org/10.5772/67463

phase. The parameter Z is assumed to be independent of time and space coordinates. From the kinetic point of view, the "correctness" of this intuitive approximation is warranted by the fact that any diffusing particle enclosed in an arbitrary cavity with inert walls evolves in time to a state where the probability to find a particle at any arbitrary point inside of a cavity does not depend on the coordinates of the particle. We call this the "filling out" of the cavity.

After the cavity has been filled, bimolecular reactions between two reactive particles enclosed in the cavity become monoexponential. In other words, a system comprised of two radicals approaches a steady state, provided by the reflection of diffusing radicals from the boundary back to the inside of the cavity. To distinguish this type of encounter from repetitive encounters in homogeneous media, we call them "enforced encounters." To illustrate this intuitive picture in more detail, we performed the RW calculations presented in **Figure 14**.

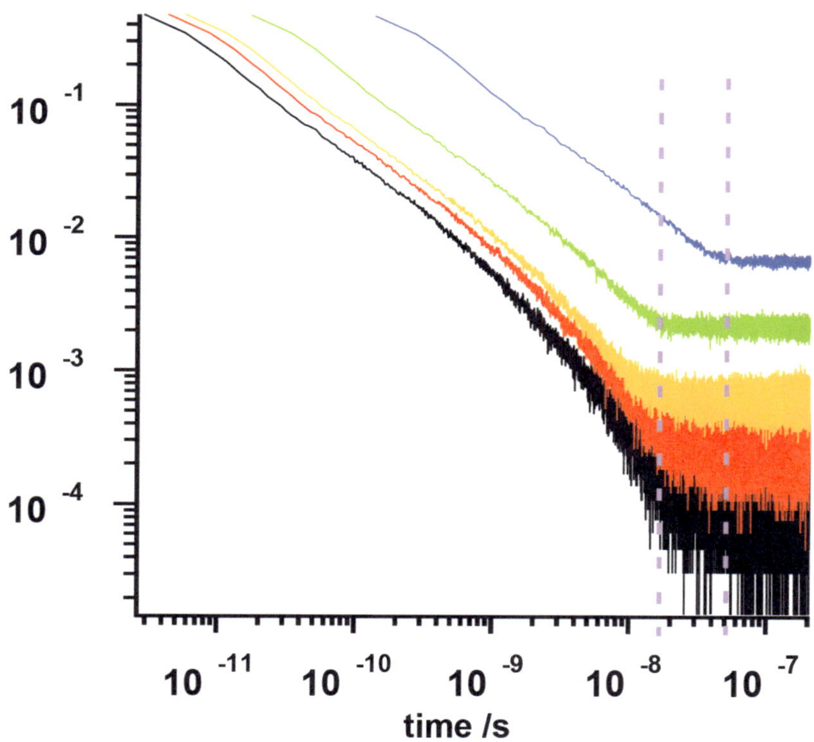

Figure 14. Filling out of AOT reverse micelles. $Lm = 53$, 33, 23, and 17×10^{-8} cm respectively for A, B, C and D traces. For the other parameters used, see section 5.4. Vertical dashed lines indicate the time of a system to arrive at the state of filled out micelle.

The probability that the next walk will be an encounter of the SCRP is shown in the plot as a function of time. Unlike our microreactor calculations, in the RW, both radicals are allowed to diffuse. In this particular calculation, the diffusion coefficients of the radicals were assumed to be the same $D_a = D_b = D$ with D taken from **Table 5**. Also, in these calculations, we propose that the SCRP is born in the center of the AOT water pool. This allows for a more direct comparison between the RW and microreactor models. From **Figure 14**, we see that indeed initially the radicals of the SCRP diffuse as if they were in unrestricted space, and the probability of their encounters decreases with time in a $t^{-3/2}$ dependence, which is a characteristic feature of unrestricted diffusion in homogeneous media. However, at some characteristic time, the kinetic behavior of the system changes dramatically: The probability of encounter stops to depend on time. The number of random walks N_L to get the boundary is equal to *ca.* $(L_m/d)^2$ where $d = 0.64 \times 10^{-8}$ cm is the length of random walk. The time at which of one of the radicals arrives at the boundary is then $t_L = (d^2/6D) \times N_L$. The time interval for these events is shown by vertical dashed lines in the figure. Thus, at the moment of the GG-AQDS radical pair observation ($\tau_w = 500$ ns), all of the reverse micelles are filled out, and the probability of encounters no longer depends on the time after SCRP creation. A consequence of this is that the observation of ST_0RPM polarization in the case of large ($L_n = 53 \times 10^{-8}$ cm) micelles, and its total absence in the small ones ($L_n = 23 \times 10^{-8}$ and 17×10^{-8} cm) is not related to the attainment of steady state conditions.

7.2. Self-quenching of the ST_0RPM CIDEP

In **Figure 15**, we present the RW calculations of the time evolution of the ST_0RPM polarization in SCRPs resulting from photoprocesses in the GG-AQDS system described in Section 4.3.5. All the parameters (except for TM polarization, which is neglected for this case) used in the calculations are given in **Table 5**. To estimate the magnitude of polarization, we calculated the population differences, being initially of zero value, between the $|\alpha\beta>$ and $|\beta\alpha>$ electron spin states of a pair with representative value 24 G for the hyperfine coupling constant (**Table 5**).

In the smallest micelles ($L_m = 17$ Å), the lifetime $\tau_D = R^2/D$ of the primary solvent cage equals 80 ns. **Figure 15** shows (black line) that strong ST_0RPM polarization has been created, but it is

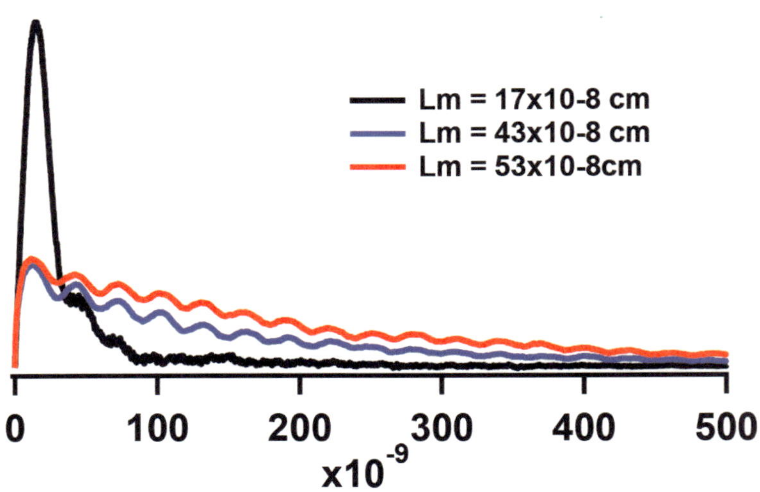

Figure 15. Generation and decay over time of the ST_0RPM polarization in AOT reverse micelles. Water core radii are given in the insertion. All other parameters are listed in **Table 4**. The TM polarization is neglected.

http://dx.doi.org/10.5772/67463

quenched during the lifetime of the primary solvent cage. This well agrees with the theoretical prediction [101] that if the geminate life is too long that $q\tau_D$ (=34) ≫ 1, then ST_0RPM polarization cannot be created. This is because the phase of the SCRP spin function becomes randomized during the encounter. In large micelles (L_m = 43 Å and 53 Å), the radicals escape the primary geminate cage being ST_0RPM polarized (blue and red lines). Thus, the RW calculations presented in **Figure 15** clearly demonstrate that it is not that the APS is transformed into the ST_0RPM but rather that the ST_0RPM polarization is created first and then is converted into the APS pattern. A question that can be asked is why the ST_0RPM polarization decreases after that

The resonance frequencies in the spin system of the SCRP are modulated by electron spin-spin Heisenberg exchange through the $\omega \pm J$ and $\pm\varepsilon$ terms (see Eq. (8) in **Table 1**). The first of these processes causes flip-flop electron spin transitions $|\alpha\beta;\chi\rangle \leftrightarrow |\beta\alpha;\chi\rangle$ which leads to annealing of the populations of corresponding spin states. The second process provides so-called "dephasing," because it splits the rotation of transverse magnetization of the RP into two components [81] and prevents the generation of ST_0RPM polarization in subsequent encounters. **Figure 16** illustrates the decay of ST_0RPM polarization due to enforced encounters. The initial state of a pair is the $|\alpha\beta\rangle$ spin state. Due to forced encounters, the exchange interaction induces flip-flop transitions in the contact pairs thereby generating a population of the $|\beta\alpha\rangle$ spin state. As follows from **Figure 16**, the system approaches to the state without any stationary polarization. Thus, the ST_0RPM polarization observed in escaped radicals is just nothing but the remnants of non-extinguished ST_0RPM polarization generated at the initial stage of SCRP life in a closed volume formed by heterogeneity, the size of which is on the nanometer scale.

We have demonstrated that the TREPR spectroscopy of SCRPs can be an effective and extremely informative tool to investigate molecular dynamics in inhomogeneous structures on the nanometer scale. The spectral shape of the APS (its components and values of spectral shifts), the contributions from ST_0RP and/or ST_RPM polarization, together with the TM

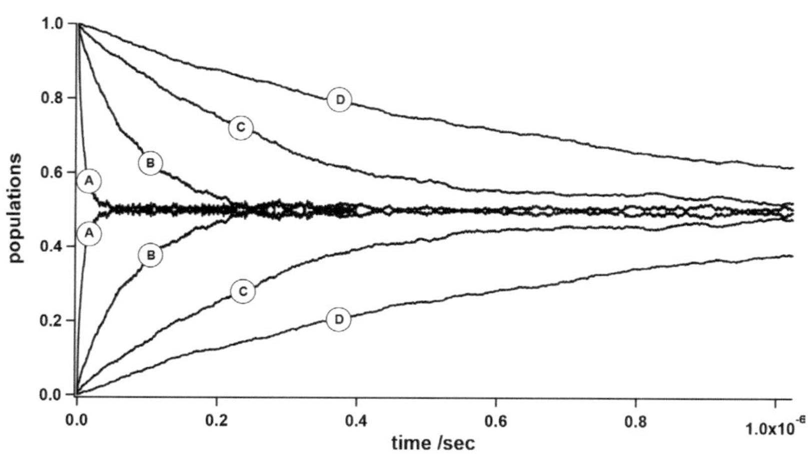

Figure 16. Relaxation of populations due to electron spin–spin exchange interaction as function of micelle size. L_m = 16 Å (A), 32 Å (B), 48 Å (C), and 64 Å (D). Both radicals diffuse, no escape, no chemical reaction, $J_0 = -2\times10^{10}$ rad/s; $\lambda = 0.5$ Å; $D_a = D_b = 10^{-6}$ cm^2.

polarization and regular encounters with rates depending on the sizes and shapes of the inhomogeneous structure provide extremely rich material for discussion, modeling, and experimentation.

The nature of the information supplied by polarized spin probes is by no means local. Rather, it is a kind of average over the volume of inhomogeneity. Through observation and simulation of the APS, we first of all demonstrate that there is no such a thing as a "location" of radicals inside inhomogeneity. If radicals were in fixed locations, then their TREPR spectra would look very different from those shown in **Figures 4**, **5**, and **8–12**.

Author details

Valery F. Tarasov[1] and Malcolm D.E. Forbes[2*]

*Address all correspondence to: forbesm@bgsu.edu

1 Semenov Institute of Chemical Physics, Moscow, Russia

2 Center for Photochemical Sciences, Department of Chemistry, Bowling Green State University, Bowling Green, Ohio, USA

References

[1] A.F. Vanin, N.A. Sanina, V.A. Serezhenkov, D.Sh. Burbaev, V.I. Losynsky, S.N. Aldoshin. Dinitrosyl–iron complexes with thiol–containing ligands: Spatial and electronic structures. Nitric Oxide Biol. Chem. **16** (2007) 62–93.

[2] G.I. Likhtenshtein "Nitroxide Spin Probes for Studies of Molecular Dynamics and Microstructure" Chapter 6, p. 205 in G.I. Likhtenshtein, J. Yamauchi, S. Nakatsuji, A. Smirnov, R. Tamura (Eds) Nitroxides: Applications in chemistry, biochemistry and materials science. Wiley – VCH Verlag GmbH & Co.KGaA (2008).

[3] A.I. Kokorin (Ed). *Nitroxides – Theory, experiment and applications.* Croatia (2012), 436 p. A free online edition of this book is available at www.intechopen.com

[4] B.H. Robinson, H. Thomann, A.H. Beth, P. Fajer, L.R. Dalton. The phenomenon of magnetic resonance: Theoretical considerations. In EPR and Advanced EPR Studies of Biological Systems. L.R. Dalton (Ed.). C.R.C.Press Boca Raton FL 1 – 314.

[5] A.L. Buchachenko, A.M. Wasserman. *Stable radicals*. Khimiya, Moscow (1979), p. 410.

[6] D.E. Budil, S. Lee, S. Saxena, J.H. Freed. Nonlinear–least–squares analysis of slow–motion EPR spectra in one and two dimensions using a modified Levenberg–Marquardt algorithm. J. Magn. Reson. A **120** (1996) 155–189.

http://dx.doi.org/10.5772/67463

[7] S. Stoll, A. Schweiger. EasySpin, a comprehensive software package for spectral simulation and analysis in EPR. J. Magn. Reson. **178** (2006) 42–55.

[8] F.H. Cho, V. Stepanov, S. Takahashi. A high-frequency electron paramagnetic resonance spectrometer for multi-dimensional, multi-frequency, and multi-phase pulsed measurements. Rev. Sci. Instrum. **85** (2014) 075110.

[9] A.J. Hoff (Ed). *Advanced EPR. Application in biology and biochemistry.* Elseiver Science Publishers B.V., Amsterdam, Oxford, New York and Tokyo (1989), 918 p.

[10] S.A. Dikanov, Yu D. Tsvetkov. *Electron spin echo envelope modulation (ESEEM) spectroscopy.* CRC Press, Boca Raton, Ann Arbor, London and Tokyo (1992), 412 p.

[11] L.J. Berliner, S.S. Eaton, G.R. Eaton (Eds). *Biological magnetic resonance. Distance measurements in biological systems by EPR,* Vol. 19. Kluwer Academic Publishers, New York, Boston, Dordrecht, London and Moscow (2002), p. 614.

[12] J. B. Feix and C. S. Klug "Site–Directed spin labeling of membrane proteins and peptide–membrane interactions" in L.J. Berliner (Ed) "Biological Magnetic Resonance, Spin Labeling: The Next Millennium, 14, Plenum Press, New York (1998) pp. 251 – 281

[13] Yu E. Nesmelov, D.D. Thomas. Protein structural dynamics revealed by site–directed spin labeling and multifrequency EPR. Biophys. Rev. **2** (2010) 91–99.

[14] D. Marsh. Spin–label EPR for determining polarity and proticity in biomolecular assemblies: Transmembrane profiles. Appl. Magn. Reson. **37** (2010) 435–454.

[15] C.F. Polnaszek, J.H. Freed. Electron spin resonance studies of anisotropic ordering, spin relaxation and slow tumbling in liquid crystalline solvents. J. Phys. Chem. **79**:2 (1975) 2283–2306.

[16] P.J. Wagner, G.S. Hammond. In A.W. Noyes, G.S. Hammond, J.N. Pitts (Eds) *Advances in photochemistry.* Wiley, Hoboken, NJ (2007), p. 5.

[17] W.M. Haynes. *CRC handbook of chemistry and physics* (94th ed.). CRC Press, Boca Raton, FL (2013).

[18] R.W. Fessenden, R.H. Schuler. Electron spin resonance studies of transient alkyl radicals. J. Chem. Phys. **39**:9 (1963) 2147–2195.

[19] J. Bargon. The discovery of chemically induced dynamic polarization (CIDNP). Helv. Chim. Acta **89** (2006) 2082–2102.

[20] H.R. Ward, R.G. Lawler. Nuclear magnetic resonance emission and enhanced absorption in rapid organometallic reactions. J. Am. Chem. Soc. **89** (1967) 5518–5519.

[21] E. Wigner, E.E. Witner. On the structure of the spectra of two–atomic molecules according to quantum mechanics. Z. Phys. **51** (1928) 859. In H. Hetema (Ed) *Quantum chemistry: Classic scientific papers.* World Scientific, Singapore, New Jersey, London and Hog Kong (2000), pp. 287–311.

[22] K.M. Salikhov, Yu N. Molin, R.Z. Sagdeev, A.L. Buchachenko. *Spin polarization and magnetic effects in radical reactions*. Akadémiai Kiadó, Budapest (1984), p. 419.

[23] H. Hayashi. *Introduction to dynamic spin chemistry. Magnetic field effects on chemical and biochemical reactions*. World Scientific Lecture and Course Notes in Chemistry, Vol. 8. S. H. Lin Editor-in-charge. World Scientific (2004).

[24] G.L. Closs. Chemically induced dynamic nuclear polarization. Adv. Magn. Reson. **7** (1974) 157–229.

[25] R. Kaptein. Chemically induced dynamic nuclear polarization: Theory and applications in mechanistic chemistry. Adv. Free Radic. Chem. **5** (1975) 319–380.

[26] U.E. Steiner, T. Ulrich. Magnetic field effects in chemical kinetics and related phenomena. Chem. Rev. **89**:1 (1989) 51–147.

[27] F.S. Sarvarov, K.M. Salikhov. Theory of spin–dependent recombination of radicals in homogeneous solution. React. Kinet. Catal. Lett. **4** (1976) 33–41.

[28] J.B. Pedersen. High field CIDNP. General analytic results. J. Chem. Phys. **67**:9 (1977) 4097–4102.

[29] L. Monchick, F.J. Adrian. On the theory of chemically induced electron polarization (CIDEP): Vector model and an asymptotic solution. J. Chem. Phys. **68** (1978) 4376–4383.

[30] F.J. Adrian. Theoretical aspects of chemically induced magnetic polarizations. Res. Chem. Intermed. **16** (1991) 99–125.

[31] J.B. Pedersen, J.H. Freed. Theory of chemically induced dynamic electron polarization. I. J. Chem. Phys. **58** (1973) 2746–2762.

[32] J.B. Pedersen, J.H. Freed. Theory of chemically induced dynamic electron polarization. II. J. Chem. Phys. **59** (1973) 2869–2885.

[33] J.B. Pedersen, J.H. Freed. Theory of chemically induced dynamic electron polarization. III. Initial triplet polarization. J. Chem. Phys. **62** (1975) 1706–1711.

[34] P.W. Atkins, G.T. Evans. Electron spin polarization in a rotating triplet. Mol. Phys. **27** (1974) 1633–1644.

[35] I.S.M. Saiful, J.–I. Fujisawa, N. Kobayashi, Y. Ohba, S. Yamauchi. A study of electron spin polarization transfer in a system of photo–excited phthalocyanine and a nitroxide radical. Bull. Chem. Soc. Jpn. **72** (1999) 661–667.

[36] M.T. Colvin, R. Carmieli, T. Miura, S. Richert, D.M. Gardner, A.L. Smeig, S.M. Dyar, S. M. Conron, M.A. Ratner, M.R. Wasielewski. Electron spin polarization transfer from photogenerated spin–correlated radical pairs to a stable radical observer spin. J. Phys. Chem. A **117** (2013) 5314–5325.

[37] S. Yamauchi. Recent developments in studies of electronic excited states by means of electron paramagnetic resonance spectroscopy. Bull. Chem. Soc. Jpn. **77** (2004) 1255–1268.

[38] A.I. Shushin. Diffusion theory of CIDEP spectra of spin correlated radical pairs. Chem. Phys. Lett. **177** (1991) 338–344.

[39] G.L. Closs, M.D.E. Forbes, J.R. Norris, Jr. Spin–polarized electron paramagnetic resonance spectra of radical pairs in micelles. Observation of electron spin–spin interactions. J. Phys. Chem. **91**:13 (1987) 3592–3599.

[40] V.R. Gorelik, V.F. Tarasov, S.R. Shakirov, E.G. Bagryanskaya. Effect of nitroxide radicals on chemically induced dynamic electron polarization of spin–correlated radical pairs in aqueous micellar solutions of sodium dodecy sulfate. Russ. Chem. Bull. Int. Ed. **57**:7 (2008) 1416–1427.

[41] A.L. Konkin, H.–K. Roth, M. Schroedner, G.A. Nazmutdinova, A.V. Aganov, T. Ida, R.R. Garipov. Time-resolved EPR study of radicals from 2,2-dimethoxy-2-phenylacetophenone in ethylene glycol after flash photolysis. Chem. Phys. **287** (2003) 377–389.

[42] A.N. Savitsky, H. Paul. Quantitative time-resolved EPR CIDEP study of the photodecomposition of *trans*-azocumene in solution. Appl. Magn. Reson. **12** (1997) 449–464.

[43] K. Hasharoni, H. Levanon, M.K. Bowman, J.R. Norris, D. Gust, T.A. Moore, A.L. Moore. Analysis of time-resolved CW_EPR spectra of short-lived radicals at different times after laser excitation. Appl. Magn. Reson. **1** (1990) 357–368.

[44] A.G. Redfield. Nuclear magnetic resonance and rotary saturation in solids. Phys. Rev. **98** (1955) 1787–1809.

[45] T.N. Makarov, H. Paul. Analysis of EPR-time profiles of transient radicals with unresolved spectra. J. Magn. Reson. **169** (2004) 335–341.

[46] V.F. Tarasov, I.S.M. Saiful, Y. Iwasaki, Y. Ohba, A. Savitsky, K. Mobius, S. Yamauchi. Electron spin polarization in an excited triplet-radical pair system: Generation and decay of the state. Appl. Magn. Reson. **30** (2006) 619–636.

[47] N.J. Turro, M.H. Kleinman, E. Karatekin. Electron spin polarization and time–resolved electron paramagnetic resonance: Applications to the paradigms of molecular and supramolecular photochemistry. Angew. Chem. Int. Ed. **39** (2000) 4436–4461.

[48] G. Kroll, M. Pluschau, K.-P. Dinse, H. van Willigen. Fourier transform–electron paramagnetic resonance spectroscopy of correlated radical pairs. J. Chem. Phys. **93**:12 (1990) 8709–8716.

[49] C.M.R. Clancy, V.F. Tarasov, M.D.E. Forbes. Time–resolved electron paramagnetic resonance studies in organic photochemistry. In B.C. Gilbert, N.M. Atherton, M.J. Davies (Eds) *Electron paramagnetic resonance, specialist periodical reports*, Vol. 16. Royal Society of Chemistry, Cambridge, UK, (1998), pp. 50–78.

[50] M.D.E. Forbes, L.E. Jaroha, S.Y. Sim, V.F. Tarasov. Time resolved spectroscopy: History, technique, and application to supramolecular and macromolecular chemistry. Adv. Phys. Org. Chem. **47** (2013) 1–83.

[51] M. Goez. An introduction to chemically induced dynamic nuclear polarization. Concepts Magn. Reson. **7** (1995) 69–86.

[52] E. Davidso, G. Jeschke, G. Matysik. Photo–CIDNP MAS NMR. In T.J. Aaartsma, J. Matysic (Eds) *Biophysical techniques in photosysnthesis II*. Springer, Dordrecht, The Netherlands (2008), pp. 385–399.

[53] K.H. Mok, P.J. Hore. Photo–CIDNP NMR methods studying protein folding. Methods **34** (2004) 75–87.

[54] E.G. Bagranskaya, R.Z. Sagdeev. Dynamic and stimulated nuclear polarization in photochemical radical reactions. Russ. Chem. Rev. **69** (2000) 925–945.

[55] K.A. McLauchlan, S.R. Nattrass. Experimantal studies of the spin–correlated radical pair in micellar and microemulsion media; MARY, RYDMR B_0 and RYDMR B_1 spectra. Mol. Phys. **65** (1988) 1483–1503.

[56] S.N. Batchelor, K.A. McLauchlan, I.A. Shkrob. Reaction yield detected magnetic resonance and magnetic field effect studies of radical pairs containing electronically excited organic radicals. Mol. Phys. **77** (1992) 75–109.

[57] J.R. Woodward, C.R. Timmel, P.J. Hore, K.A. McLauchlan. Low field RYDMR: Effects of orthogonal static and oscillating magnetic fields on radical recombination reactions. Mol. Phys. **100** (2002) 1181–1186.

[58] S. Aich, S. Basu. Magnetic field effect: A tool for identification of spin state in a photoinduced electron–transfer reaction. J. Phys. Chem. A **102** (1998) 722–729.

[59] K. Schulten. Magnetic field effects in chemistry and biology. In J. Treusch (Ed) *Festrorperproblem*. Advances in solid state physics, Vol. XXII. Vieweg, Braunschweig (1982), pp. 61–83.

[60] C.B. Grisom. Magnetic field effects in biology: A survey of possible mechanisms with emphasis on radical–pair recombination. Chem. Rev. **95** (1995) 3–24.

[61] V.F. Tarasov, N.D. Ghatlia, N.I. Avdievich, I.A. Shkrob, A.L. Buchachenko, N.J. Turro. Examination of the exchange interaction through micelle size. 2. Isotope separation efficiency as an experimental probe. J. Am. Chem. Soc. **116** (1994) 2281–2291.

[62] A.L. Buchachenko. Magnetic isotope effect: Nuclear spin control of chemical reactions. J. Phys. Chem. A **105** (2001) 9995–10011.

[63] A.R.P. Rau. Manipulating two–spin coherences and qubit pairs. Phys. Rev. A **61** (2000) 032301.

[64] X. Hu, S.D. Sarma. Hilbert–space structure of a solid–state quantum computer: Two electron states of a double–quantum–dot artificial molecule. Phys. Rev. A **61** (2000) 062301.

[65] Y. Omar, N. Paunkovic, S. Bose, V. Vedral. Spin–space entanglement transfer and quantum statistics. arXiv:quant–ph/0105120 **2** (2002) 5.

[66] D.P. DiVincenzo. Real and realistic quantum computers. Nature **393** (1998) 113–114.

[67] S.D. Sarma. Spintronics. A new class of device based on electron spin, rather than on charge, may yield the next generation of microelectronics. Am. Sci. **89** (2001) 516–523.

[68] C.D. Buckley, D.A. Hunter, P.J. Hore, K.A. McLauchlan. Electron spin resonance of spin–correlated radical pairs. Chem. Phys. Lett. **135** (1987) 307–311.

[69] Y. Sakaguchi, H. Hayashi, H. Murai, Y.J. I'Haya. CIDEP study of the photochemical reactions of carbonyl compounds showing the external magnetic field effect in a micelle. Chem. Phys. Lett. **110** (1984) 275–279.

[70] Y. Sakaguchi, H. Hayashi, H. Murai, Y.J. I'Haya, K. Mochida. CIDEP study of the formation of a cyclohexadienyl–type radical in the hydrogen abstraction reactions of triplet xanthone. Chem. Phys. Lett. **120** (1985) 401–405.

[71] H. Murai, Y. Sakaguchi, H. Hayashi, Y.J. I'Haya. An anomalous phase effect in the individual hyperfine lines of the CIDEP spectra observed in the photochemical reactions of benzophenone in micelles. J. Phys. Chem. **90** (1986) 113–118.

[72] A.D. Trifunac, D.J. Nelsen. Chemically induced dynamic electron polarization. Pulse radiolysis of aqueous solutions of micelles. Chem. Phys. Lett. **46** (1977) 346–348.

[73] G.L. Closs, M.D.E. Forbes, J.R. Norris, Jr. Spin–polarized electron paramagnetic resonance spectra of radical pairs in micelles. Observation of electron spin–spin interactions. J. Phys. Chem. **91** (1987) 3592–3599.

[74] S.H. Glarum, J.H. Marshall. Spin exchange in nitroxide biradicals. J. Chem. Phys. **47** (1967) 1374–1378.

[75] H. Yonemura, M.D.E. Forbes. Electron spin exchange in linked phenothiazine–viologen charge transfer complexes incorporated in "through–ring" (rotaxane) α–cyclodextrins. Photochem Photobiol. **91** (2015) 672–677.

[76] P. Caregnato, L.E. Jarocha, H.S. Esinhart, N.V. Lebedeva, V.F. Tarasov, M.D.E. Forbes. Electrostatic control of spin exchange between mobile spin–correlated radical pairs created in micellar solutions. Langmuir **112** (2011) 7574–7580.

[77] A. Kawai, A. Shikama, M. Mutsui, K. Obi. Time resolved ESR study on the photochemical reactions of pyrene and nitroxide radical system in micelle; formation of spin–correlated pyrene cation – nitroxide pairs. Bull. Chem. Soc. Jpn. **74** (2001) 1203–1211.

[78] E.E. Chaney, M.D.E. Forbes. Dynamics of spin–correlated radical pairs in non–ionic surfactant solutions. J. Phys. Chem. B **107** (2003) 4464–4469.

[79] S. Moribe, T. Ikoma, K. Akiyama, S. Tero–Kubota. Time–resolved EPR study on photoreduction of sodium anthraquinone–2–sulfate in liposome. Chem. Phys. Lett. **457** (2008) 66–68.

[80] Q. Mi, M.A. Ratner, M.R. Wasielewski. Time–resolved EPR spectra of spin–correlated radical pairs: Spectral and kinetic modulation resulting from electron–nuclear hyperfine interactions. J. Phys. Chem. A **114** (2010) 162–171.

[81] V.F. Tarasov, M.D.E. Forbes. Time resolved electron spin resonance of spin correlated micelle confined radical pairs. Shape of the anti–phase structure. Spectrochim. Acta A **56** (2000) 245–263.

[82] V.F. Tarasov, L.E. Jarocha, N.I. Avdievich, M.D. Forbes. TREPR spectra of micelle – confined spin correlated radical pairs. I. Molecular motion and simulations. Photochem. Photobiol. Sci. **13** (2014) 439–453.

[83] C.–H. Wu, W.S. Jenks, I.V. Koptyug, N.D. Ghatlia, M. Lipson, V.F. Tarasov, N.J. Turro. Time–resolved ESR examination of a simple supramolecular guest–host system. Electron spin exchange interaction in micellized spin–correlated radical pairs. J. Am. Chem. Soc. **115** (1993) 9583–9595.

[84] V.F. Tarasov, H. Yashiro, K. Maeda, T. Azumi, I.A. Shkrob. Spin–correlated radical pairs in micellar systems: Mechanism of CIDEP and the micelle size dependence. Chem. Phys. **212** (1996) 353–361.

[85] V.F. Tarasov, H. Yashiro, K. Maeda, T. Azumi, I.A. Shkrob. Time–resolved ESR in a spin–correlated radical pair with large hyperfine coupling constant at ^{31}P. Micellar size effects and the role of flip–flop transitions. Chem. Phys. **226** (1998) 253–269.

[86] K. Maeda, M. Terazima, T. Azumi, Y. Tanimoto. CIDNP and CIDEP studies on intramolecular hydrogen abstraction reaction of polymethylene–linked xanthone and xanthene. Determination of the exchange integral of the intermediate biradicals. J. Phys. Chem. **95** (1991) 197–204.

[87] K. Tominaga, S. Yamauchi and N. Hirota. A chemically induced dynamic electron polarization study on the acetone ketyl radical and radical pair in an alcohol solution. J. Chem. Phys. **92** (1990) 5175–5185.

[88] A.I. Shushin. The cage effect and ESR spectra of spin–correlated radical pairs. Chem. Phys. Lett. **162** (1989) 409–415.

[89] L. Sterna, D. Ronis, S. Wolfe, A. Pines. Viscosity and temperature dependence of the magnetic isotope effect. J. Chem. Phys. **73**:11 (1980) 5493–5499.

[90] V.F. Tarasov, A.L. Buchachenko, V.I. Maltsev. Magnetic isotope effect and isotope separation in microreactors. Russ. J. Phys. Chem. **55** (1981) 1921–1928.

[91] V.F. Tarasov, N.D. Ghatlia, N.I. Avdievich, I.A. Shkrob, A.L. Buchachenko, N.J. Turro. Examination of the exchange interaction through micelle size. 2. Isotope separation efficiency as an experimental probe. J. Am. Chem. Soc. **116** (1994) 2281–2291.

[92] N.J. Turro, G.S. Cox, M.A. Paczkowski. Photochemistry in micelles. Top. Curr. Chem. **129** (1985) 57–97.

[93] M. Tachiya. Stochastic and diffusion models of reactions in micelles and vesicles. In G.R. Freeman (Ed) Kinetics of nonhomoheneous processes. A practical introduction for chemists, biologists, physicists, and materials scientists. John Wiley & Sons, Inc., New York, NY, (1987) 575–650

[94] V.F. Tarasov, N.D. Ghatlia, A.L. Buchachenko, N.J. Turro. Probing the exchange interaction through the micelle size. 1. Probability of recombination of triplet geminate radical pairs. J. Am. Chem. Soc. **114** (1992) 9517–9526.

[95] R.C. White, V.F. Tarasov, M.D.E. Forbes. Photooxidation of diglycine in confined media. Application of the microreactor model for spin–correlated radical pairs in reverse micelles and water–in–oil microemulsions. Langmuir **21** (2005) 2721–2727.

[96] V.F. Tarasov, L.E. Jaroha, M.D. Forbes. TREPR spectra of micelle – confined spin correlated radical pairs. II. Spectral decomposition and asymmetric line shapes. Photochem. Photobiol. Sci. **13**:2 (2014) 454–463.

[97] Y. Sakaguchi, H. Hayashi, H. Murai, Y.J. I'Haya. CIDEP study of the photochemical reactions of carbonyl compounds showing the external magnetic field effect in a micelle. Chem. Phys. Lett. **110** (1984) 275–279.

[98] M. Hasegawa, T. Sugimura, Y. Suzaki, Y. Suzaki, Y. Shindo. Microviscosity in water pool of aerosol–OT reversed micelle determined with viscosity–sensitive fluorescenck probe, auramine 0, and fluorescence depolarization of xanthene dyes. J. Phys. Chem. **98** (1994) 2120–2124.

[99] K.L. Aminov, J.B. Pedersen. Improved finite difference calculation of magnetic field effects: Inclusion of analytic asymptotics. Chem. Phys. **193** (1995) 297–308.

[100] H.–X. Zhou. Comparison of three Brownian–dynamics algorithms for calculating rate constants of diffusion–influenced reactions. J. Chem. Phys. **108** (1998) 8139–8145.

[101] J.H. Freed, J.B. Pedersen. Theory of chemically induced dynamic spin polarization. Adv. Magn. Reson. **8** (1976) 1–84.

Chapter 6

Thermodynamics and Bending Energetics of Microemulsions

L. Magnus Bergström

Additional information is available at the end of the chapter

http://dx.doi.org/10.5772/67369

Abstract

A comprehensive, yet simple, theoretical model for droplet microemulsions is presented. The model combines thermodynamics of self-assembly with bending elasticity theory and relates microemulsion properties, such as average droplet size, polydispersity, interfacial tension and solubilisation capacity with the three bending elasticity constants, spontaneous curvature (H_0), bending rigidity (k_c) and saddle-splay constant ($\bar{k}_c$). In addition, the self-association entropy constant (k_s) explicitly determines various microemulsion properties. The average droplet size is shown to increase with increasing effective bending constant, defined as $k_{eff} = 2k_c + \bar{k}_c + k_s$, as well as with decreasing magnitudes of H_0. The polydispersity decreases with increasing values of k_{eff}, but does not at all depend on H_0. The model predicts ultra-low interfacial tensions, the values of which decrease considerably with increasing droplet radius, in agreement with experiments. The solubilisation capacity increases as the number of droplets is increased with increasing surfactant concentration. In addition, an enhanced solubilisation effect is obtained as the size of the droplets increases with increasing surfactant concentration, as a result of self-association entropy effects. It is demonstrated that self-association entropy effects favour smaller droplet size as well as larger droplet polydispersity.

Keywords: surfactant, self-assembly, microemulsion, interfacial tension, solubilisation, spontaneous curvature, bending rigidity, saddle-splay constant

1. Introduction

Surfactants are amphiphilic molecules consisting of a hydrophilic head group and a hydrophobic tail. Surfactants self-assemble above a certain surfactant concentration in an aqueous solvent to form micelles or bilayers. The driving force for the self-assembly process is the tendency to minimize the interfacial contact area between water and surfactant hydrophobic tails. As a result, the tails make up an interior core of the micelles that is absent of water whereas the hydrophilic head groups are located at the interface adjacent to the aqueous solvent. Hydrophobic components that are usually insoluble in water, such as oil or fat, may be dissolved into a surfactant-water system. The oil molecules become incorporated into the hydrophobic core of the micelles to form swollen micelles. However, there is a limit how much oil that may be dissolved by micelles and above a certain amount of oil two separate liquid phases coexist, i.e. thermodynamically stable oil droplets (discrete phase) dissolved in water (continuous phase) coexisting with excess oil (cf. **Figure 1**). Such an oil-in-water (o/w) microemulsion phase coexisting with an excess oil phase in a two-phase system is denoted as Winsor I microemulsion. Some surfactants, on the other hand, may form reversed water-in-oil (w/o) droplets that dissolve water (discrete phase) in oil (continuous phase) in the presence of excess water (Winsor II microemulsion). Moreover, a microemulsion phase may also coexist with both excess oil and excess water in a three-phase system (Winsor III microemulsion) [1]. A Winsor III microemulsion does not usually consist of finite-sized droplets. Rather some kind of macroscopic structure is formed, either more or less planar alternating layers of oil and water separated by surfactant monolayers, or some kind of ordered or disordered bicontinuous phase with a system of separated tunnels of water and oil [2].

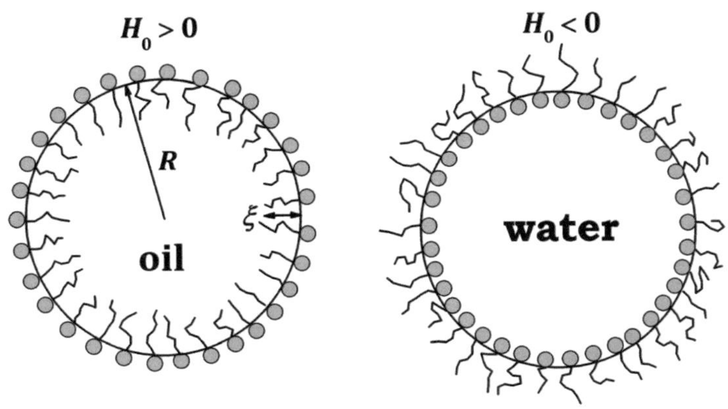

Figure 1. Schematic figures showing a spherical oil-in-water (o/w) microemulsion droplet (left) and a water-in-oil (w/o) microemulsion droplet (right). The radius of the droplet is denoted R and the thickness of the surfactant monolayer at the interface is denoted ξ. Oil-in-water droplets are formed by surfactants with a positive spontaneous curvature ($H_0 > 0$) whereas w/o droplets are formed by surfactants with $H_0 < 0$.

The microscopic structure of microemulsions (whether Winsor I, II or III), is mainly determined by the chemical structure of the surfactant. Surfactants with large hydrophilic head groups and small hydrophobic (or lipophilic if one prefers) parts tend to curve so as to form ordinary micelles or o/w microemulsion droplets whereas surfactants with a small hydrophilic part and large hydrophobic part (like phospholipids) tend to form w/o microemulsion droplets (cf. **Figure 1**). The molecular properties determining the curvature of a surfactant monolayer may be summed up in the quantity hydrophilic-lipophilic balance, or shorter the HLB value [3]. As a result, one expects o/w microemulsion droplets to increase in size as the monolayer becomes less curved with decreasing HLB values. For sufficiently low HLB values, a transition from Winsor I to Winsor III is expected, and further decreasing HLB would result in the formation of Winsor II microemulsion. In addition to the volume of the hydrophilic and hydrophobic parts, respectively, the presence of electric charge on the surfactant head group also contributes to increase the HLB value and promotes a positive (oil-in-water) curvature of the droplets.

It is possible to tune the HLB value of ionic surfactants by means of adding salt to an aqueous phase and HLB decreases with increasing electrolyte concentration. Similarly, the HLB value of non-ionic ethylene oxide-based surfactants is found to be considerably temperature-sensitive and, as a result, it is possible to observe the sequence of transitions oil-in-water → bicontinuous → water-in-oil microemulsion by means of increasing the temperature [4]. The amount of oil dissolved in the microemulsion increases during this sequence of transitions and, at a certain temperature denoted the phase inversion temperature (PIT), the microemulsion phase contains equal amounts of oil and water. It is also possible to tune the curvature of a surfactant monolayer by means of adding a cosurfactant that is mixed into the layer [5].

In contrast to conventional (macro)emulsions, microemulsions are thermodynamically stable systems. This means that the size of the droplets may fall within a wide range, depending on the surfactant HLB value, from small micelle-like droplets of about 1 nm to about 100 nm, above the size of which the droplets usually transform into a macroscopic bicontinuous structure. In contrast, kinetically stabilized emulsion droplets are usually larger falling in the range 10 nm to 100 μm.

In this chapter, we present a comprehensive, yet simple, theory that rationalizes the structural behaviour of spherical microemulsion droplets. The theory is based on conventional solution thermodynamics combined with bending elasticity theory and it predicts several experimentally available quantities, such as droplet size and polydispersity, interfacial tension and solubilisation capacity. We only consider the case of rather rigid microemulsion droplets that are spherically shaped. More flexible droplets, consisting of interfacial monolayers with bending rigidities approaching zero (see further below), may assume a more spheroidal or ellipsoidal shape [6] or undergo undulatory fluctuations of the droplet interfaces [7].

2. Thermodynamics of self-assembly

Microemulsions are thermodynamically stable equilibrium structures. Hence, the theoretical treatment of microemulsion droplets necessarily needs to take into account thermodynamics of self-assembling surfactant molecules to form an interfacial monolayer that encapsulates the discrete phase (cf. **Figure 1**). The latter process may be considered in terms of a set of multiple equilibrium reactions

$$N \text{ Surfactant} \rightleftharpoons M_N \tag{1}$$

where N is the free surfactant molecules dissolved in the continuous phase (Bancroft's rule states that free surfactants are always present in the continuous phase) that form a microemulsion droplet M_N. The change in entropy for the process of self-assembling free surfactant molecules to the interface of a single droplet that encloses a certain amount of oil (o/w microemulsions) or water (w/o microemulsions) in accordance with Eq. (1) may thus be written as

$$\Delta S_{agg} = -k(\ln\phi_N - N\ln\phi_{free}) \tag{2}$$

where k is the Boltzmann's constant. For the case of o/w microemulsions, the free surfactant molecules are present in the aqueous continuous phase with a volume fraction ϕ_{free}, whereas oil molecules are brought from the adjacent macroscopic excess oil phase to form droplets encapsulated by surfactant monolayers with volume fraction ϕ_N. The expression in Eq. (2) is also valid for the reverse case of w/o microemulsions in which the surfactant is dissolved in the continuous oil phase and water is brought from an excess aqueous phase. ΔS_{agg} is always a negative quantity and, as a result, the process of self-assembly must be unfavourable in the absence of other contributions favouring the self-assembly process. Equation (2) can be understood as the difference between the entropy of mixing self-assembled surfactant monolayers and solvent, and the entropy of mixing the corresponding amount of free surfactant molecules with solvent.

In order to allow for the spontaneous self-assembly of surfactant molecules, additional contributions to the overall free energy change must be added. These contributions are related to the formation of the surfactant monolayer and can be collected into a single quantity denoted by $\Delta\mu$. The most important contribution to $\Delta\mu$ comes from the hydrophobic effect, i.e. the largely entropic favourable process of reducing interfacial contact between surfactant hydrophobic tails and water molecules as free surfactants are self-assembled in the microemulsion-droplet interfaces. From a molecular point of view, the hydrophobic effect is believed to be the result of organization of water molecules adjacent to the hydrophobic moieties of the surfactant molecules. We may now write the total change in Gibbs free energy of the self-assembly process as

$$\Delta G = N\Delta\mu - T\Delta S_{agg} \tag{3}$$

Introducing the following free energy parameter [8].

$$\mathrm{E}_N \equiv N\Delta\mu - NkT\ln\phi_{free} \tag{4}$$

and combining Eqs. (2)–(4) gives the following set of equilibrium conditions

$$\Delta G_N = \mathrm{E}_N + kT\ln\phi_N = 0 \tag{5}$$

one for each surfactant aggregation number N of the interfacial monolayers. By setting $\Delta G_N = 0$, we account for a situation where surfactant molecules are reversibly exchanged between droplet interface and as free molecules in the continuous phase. Equation (5) may now be rewritten as

$$\phi_N = e^{-\mathrm{E}_N/kT} \tag{6}$$

Summing up the different volume fractions in Eq. (6) gives the total volume fraction ϕ_s of surfactant self-assembled in the droplets, i.e.

$$\phi_s = \sum_{N=2}^{\infty} \phi_N \quad (7)$$

which may be evaluated from the following integral approximation

$$\phi_s = \int_0^{\infty} e^{-E(N)/kT}\, dN \quad (8)$$

We have been able to approximately set the lower limit in the integral to zero since the largely curved droplets with low aggregation numbers are too energetically unfavourable to contribute to the integral (see further below). Equation (8) gives the full size distribution of surfactants self-assembled in microemulsion-droplet interfaces; in so far an mathematical expression for the function E(N) is available. In the following section, we will demonstrate that such an expression may be obtained by means of considering the properties of bending a surfactant monolayer.

3. Bending elasticity

The free energy of an arbitrarily shaped surfactant monolayer may be calculated taking into account bending elasticity properties. The curvature at a single point on a surfactant monolayer (most conveniently defined at the hydrocarbon/water interface [cf. **Figure 1**]) may be defined by considering two perpendicular curves on the interface with radii of curvature, R_1 and R_2, respectively. Hence, each point at the aggregate interface may be distinguished by two principal curvatures defined as $c_1 = 1/R_1$ and $c_2 = 1/R_2$. By means of changing variables from c_1 and c_2 to the mean curvature $H = ½(c_1 + c_2)$ and Gaussian curvature $K = c_1 \cdot c_2$, respectively, the following fruitful expression for the free energy per unit area may be introduced [9]

$$\gamma(H,K) = \gamma_0 + 2k_c(H - H_0)^2 + \bar{k}_c K \quad (9)$$

The Helfrich expression in Eq. (9) is a second-order expansion with respect to H and K, respectively, and introduces three important parameters, spontaneous curvature (H_0), bending rigidity (k_c) and saddle-splay constant ($\bar{k}_c$). γ_0 is the free energy per unit area at $H = H_0$ and $K = 0$. The total free energy of a surfactant monolayer may be obtained by means of integrating Eq. (9) over the entire interfacial area A of the monolayer giving

$$E = \int \gamma(H,K)dA = \gamma_0 A + 2k_c \int (H - H_0)^2 dA + \bar{k}_c \int K dA \quad (10)$$

For a spherically curved monolayer, an expression for E is obtained by simply introducing the proper values $H = 1/R$ and $K = 1/R^2$ in Eq. (9) and multiplying γ with the interfacial area $4\pi R^2$. As a result, the following expression for E as a function of droplet radius R is obtained for a surfactant monolayer at the interface of a spherical microemulsion droplet

$$E(R) = 4\pi(2k_c + \bar{k}_c - 4k_c H_0 R + \gamma_p R^2) \quad (11)$$

where γ_p is the interfacial tension of a strictly planar interface, i.e. $\gamma_p = \gamma(H = K = 0)$. The free energy in Eq. (11) may be considered as a sum of the contribution of forming a planar surfactant monolayer out of free surfactant ($=4\pi R^2\gamma_p$) and the contribution of bending the monolayer into spherical shape.

The three bending elasticity constants H_0, k_c and $\bar{k}_c$ influence various properties of self-assembled surfactant monolayers and are expected to assume values that reflect the chemical properties of the constituent surfactant or surfactant mixture as well as environmental properties such as temperature and electrolyte concentration. They may be interpreted as thermodynamic parameters that may be determined from experiments [10–13] as well as calculated from a suitable molecular model [14–16] by means of minimizing the free energy per molecule of a surfactant interfacial layer at given values of H and K.

3.1. Spontaneous curvature

The spontaneous curvature H_0 represents the sign and magnitude of the preferential curvature of a single surfactant monolayer. Since smaller microemulsion droplets must be more curved than larger ones, H_0 is expected to have a large impact on the droplet size. We have defined the mean and spontaneous curvature of o/w microemulsion droplets as positive, whereas w/o droplets have a negative spontaneous curvature (cf. **Figure 1**). H_0 depends on the chemical structure of a surfactant molecule in so far it is expected to increase with an increasing HLB value [14, 16]. This means that H_0 rapidly decreases as the tail length is increased for a surfactant with given head group. Likewise, ionic surfactants usually have larger spontaneous curvature than non-ionic surfactants with a head group of similar size, and adding electrolyte is expected to reduce H_0 of an ionic surfactant monolayer. As a result, in the presence of both water and oil, surfactants with high HLB values and positive H_0 tend to form o/w microemulsion droplets whereas surfactants with low HLB values and negative H_0 tend to form w/o microemulsions.

3.2. Bending rigidity

The bending rigidity k_c is a measure of the ability of a surfactant monolayer to resist deviations from a uniform mean curvature $H = H_0$. k_c must always be a positive quantity in order to realize a minimum of γ as a function of mean curvature in the Helfrich expression in Eq. (9). Large bending rigidities are expected to favour surfactant monolayers with small deviations from a homogenous curvature, i.e. rigid and monodisperse droplets with a uniform shape. This means that microemulsion droplets are expected to be strictly spherical for sufficiently large values of k_c. Likewise, a bicontinuous microemulsion is expected to form an ordered cubic phase as the bending rigidity is high whereas a more disordered sponge phase forms at low k_c values [17].

http://dx.doi.org/10.5772/67369

From a molecular point of view, it has been demonstrated that k_c assumes a maximum value for some optimal value of the HLB of a surfactant [16]. The flexibility of the surfactant tails appears to have a crucial impact on k_c which is found to be significantly larger for surfactants with a flexible rather than rigid tail [14]. As a matter of fact, it has been demonstrated that k_c must always be equal to zero, thus preventing self-assembly of non-ionic surfactants made up of hydrophobic and hydrophilic parts that both are rigid [16]. For surfactants with a flexible tail, the bending rigidity is expected to increase with increasing tail length [10, 14]. Moreover, k_c has been found to become substantially reduced in surfactant mixtures [18–20] and the magnitude of the reduction has been demonstrated to increase with increasing asymmetry between two surfactants with respect to head group charge number, volume of surfactant tail, etc [15].

3.3. Saddle-splay constant

The saddle-splay constant $\bar{k}_c$ is related to the Gaussian curvature K. As implied by its name, high positive values of $\bar{k}_c$ influence the curvature of an interface so as to favour a saddle-like structure, with negative Gaussian curvature, that is curved in opposite directions perpendicular to one another. Examples of surfactant assemblies with saddle-like structure are torus-like micelles and bicontinuous microemulsions. Hence, the latter is expected to be favoured by large (positive) values of $\bar{k}_c$ whereas a planar lamellar Winsor III microemulsion is favoured by low and negative values of $\bar{k}_c$ [17].

Moreover, according to the Gauss-Bonnet theorem, the last integral in Eq. (10) equals

$$\int K dA = 4\pi(1-g) \tag{12}$$

where g represents the number of handles or holes present in a surfactant monolayer. A microemulsion droplet consists of one geometrically closed monolayer and, as a result, $g = 0$. As a consequence, the last term in Eq. (10) equals $4\pi\bar{k}_c$, the quantity of which does not depend on the size of the microemulsion droplet. Since $\bar{k}_c$ contributes with a size-independent term to E, the value of $\bar{k}_c$ indirectly determines the size of the droplets. Positive values of $\bar{k}_c$ indicate that the total free energy in the system increases as one single droplet is split up to form two or many smaller droplets. Likewise, negative values of $\bar{k}_c$ imply a decreasing total free energy as droplets are split up to increase their number while maintaining the surfactant concentration fixed. In either case, increasing values of $\bar{k}_c$ favour larger microemulsion droplets.

From a molecular point of view, $\bar{k}_c$ is expected to be negative for a pure symmetric surfactant that tends to curve equivalently in all lateral directions [14]. However, introducing some kind of asymmetry of the surfactant may result in positive values of $\bar{k}_c$. For instance, the non-uniform distribution of charges in monolayers formed by ionic gemini surfactants may result in a positive saddle-splay constant [13, 21]. Likewise, mixtures of two or more asymmetric surfactants, in particular mixtures of oppositely charged surfactants, may result in positive values of $\bar{k}_c$ [22].

4. Microemulsion droplets

The size distribution of surfactant monolayers in Eq. (8) may be rewritten so as to give

$$\phi_s = \int_0^\infty \frac{dN}{dR} e^{-E(R)/kT} dR = \int_0^\infty \phi_s(R) dR \tag{13}$$

where $\phi_s(R)$ is the volume fraction distribution of surfactant monolayers and

$$\frac{dN}{dR} = \frac{8\pi R \xi}{v} \tag{14}$$

may be evaluated from the geometrical relation $Nv = 4\pi R^2 \xi$ for a surfactant monolayer with thickness ξ and surfactant molecular volume v. However, the self-assembled monolayer is excluded from the interior volume occupied by the discrete phase of oil molecules (in the case of o/w droplets). As a result, it is more appropriate to consider the size distribution based on the volume fraction of entire droplets ϕ_d, including both the surfactant monolayers at the droplet interfaces and solubilised oil molecules in the interior core of the droplets.

The geometrical relations $\phi_d(R) \propto 4\pi R^2 dR$ and $\phi_s(R) \propto 8\pi R \xi dR$ imply that the volume fraction distribution of droplets equals

$$\phi_d(R) = \frac{R}{2\xi} \phi_s(R) \tag{15}$$

Now we may substitute Eq. (11) into Eq. (13) and combine with Eqs. (14) and (15) to arrive at the full size distribution of spherical microemulsion droplets

$$\phi_d(R) = \frac{4\pi R^2}{v} e^{-4\pi(2k_c + \bar{k}_c - 4k_c H_0 R + \gamma_p R^2)/kT} \tag{16}$$

from which the total volume fraction of droplets (including both surfactant and the discrete phase) may be evaluated so as to give

$$\phi_d = \frac{4\pi}{v} e^{-4\pi(2k_c + \bar{k}_c)/kT} \int_0^\infty R^2 e^{-4\pi(\gamma_p R^2 - 4k_c H_0 R)/kT} dR \tag{17}$$

The size distribution according to Eq. (16) is plotted in **Figure 2** for some different values of volume fraction of droplets. In accordance with Eq. (16), the formation of finite-sized microemulsion droplets may be considered as the result of bending elasticity properties as well as self-association entropy effects. The latter contribution always favours small droplets. Since the driving force towards smaller droplets due to the entropy of self-assembly increases in magnitude with decreasing droplet volume fraction, the average size of the droplets depends on ϕ_d and the droplets increase in size with increasing surfactant (or droplet) concentration [cf. further below].

Equation (17) may be solved so as to give

http://dx.doi.org/10.5772/67369

$$\phi_d = \frac{k_{eff}(kT)^2}{128\pi v(k_c H_0)^3} e^{4\pi k_s/kT}\left(\sqrt{\pi\alpha}(2\alpha+1)(1+erf\sqrt{\alpha})+2\alpha e^{-\alpha}\right) \tag{18}$$

where we have introduced the dimensionless parameter [6, 23]

$$\alpha = \frac{16\pi(k_c H_0)^2}{\gamma_p kT} = \frac{4\pi}{kT}(2k_c + \bar{k}_c + k_s) \tag{19}$$

as well as the quantity

$$k_s = \frac{kT}{4\pi}\ln\left(\frac{128\pi v(k_c H_0)^3\phi_d}{k_{eff}(kT)^2\left(\sqrt{\pi\alpha}(2\alpha+1)(1+erf\sqrt{\alpha})+2\alpha e^{-\alpha}\right)}\right) \tag{20}$$

Except for values of α close to zero, the following two approximations of Eqs. (18) and (20), respectively, are accurate

$$\phi_d = \frac{\pi k_{eff}^{5/2}(kT)^{1/2}}{4v(k_c H_0)^3} e^{4\pi k_s/kT} \tag{21}$$

$$k_s = \frac{kT}{4\pi}\ln\left(\frac{4v(k_c H_0)^3\phi_d}{\pi k_{eff}^{5/2}(kT)^{1/2}}\right) \tag{22}$$

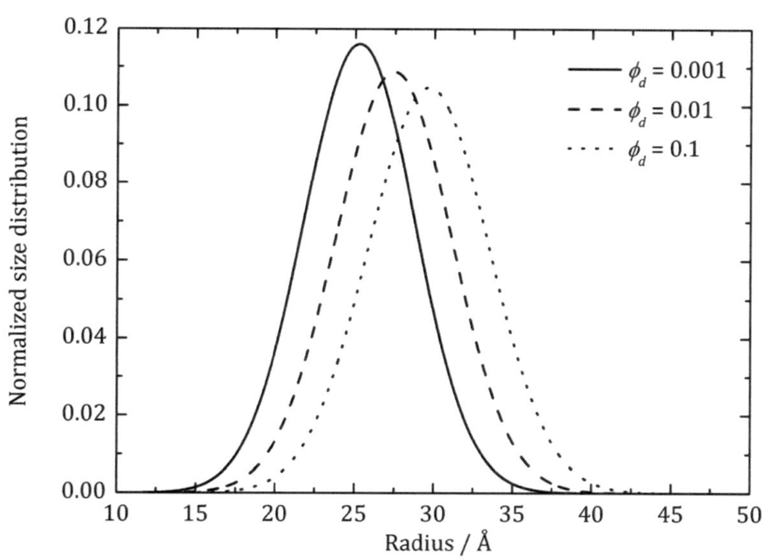

Figure 2. The normalized size distributions of microemulsion droplets at some different values of the droplet volume fraction ϕ_d. The bending elasticity constants were set to $k_c = 1.0kT$, $\bar{k}_c = 0.725kT$ and $\xi H_0 = 0.645$, where $\xi = 17$ Å, as obtained for a gemini surfactant forming elongated micelles in water [13]. The average droplet radii are seen to be always smaller than those expected from the expression $\langle R\rangle = (2k_c + \bar{k}_c)/2k_c H_0 = 36$ Å.

The entropy parameter k_s may be added to the sum of twice the bending rigidity and the saddle-splay constant to make up the effective bending constant $k_{eff} = 2k_c + \bar{k}_c + k_s$. Hence, k_s has a similar impact on microemulsion droplet properties, such as average size and polydispersity (see further below), as k_c and $\bar{k}_c$, although it is not (primarily) related to bending elasticity properties. From a physical point of view, it is a measure of the entropy of self-assembly and, since this process is unfavourable from an entropic point of view, k_s is always found to be a negative property (cf. **Figure 3**). Moreover, the entropy of self-assembly increases in magnitude as the concentration of surfactant decreases. As a result, k_s becomes a strong function of volume fraction of surfactant and its magnitude increases (that is k_s becomes more negative) as a solution of microemulsion droplets is diluted.

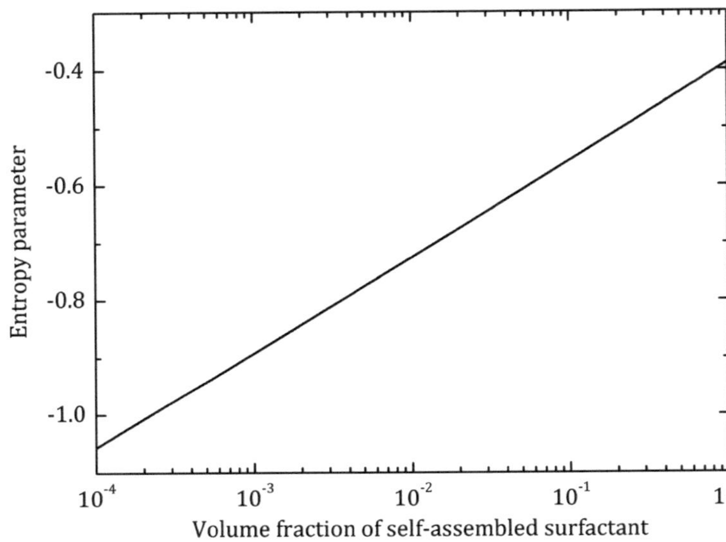

Figure 3. The self-assembly entropy parameter k_s plotted against the volume fraction ϕ_s of self-assembled surfactant, in accordance with Eq. (20). The three bending elasticity constants k_c, $\bar{k}_c$ and H_0 assume identical values as in **Figure 2**.

5. Average droplet size

The following expression for the (volume weighted) average droplet radius may be derived from the size distribution in Eq. (16)

$$\langle R \rangle = \frac{1}{\phi_d}\int_0^\infty R\phi_d dR = \frac{k_{eff}}{2k_c H_0}\frac{1}{x} \tag{23}$$

where the quantity

$$x = \frac{\sqrt{\pi\alpha}(2\alpha+1)(1+erf\sqrt{\alpha}) + 2\alpha e^{-\alpha}}{\sqrt{\pi\alpha}(2\alpha+3)(1+erf\sqrt{\alpha}) + 2\alpha e^{-\alpha} + 2e^{-\alpha}} \tag{24}$$

only depends on α. For sufficiently large values of α, Eq. (24) may be simplified so as to give

$$\frac{1}{x} = \frac{2\alpha + 3}{2\alpha + 1} = 1 + \frac{2}{2\alpha + 1} = 1 + \frac{2}{8\pi k_{eff}/kT + 1} \tag{25}$$

Combining Eqs. (23) and (25) gives the following rather simple expression for the average droplet radius

$$\langle R \rangle = \frac{k_{eff}}{2k_c H_0}\left(1 + \frac{2}{8\pi k_{eff}/kT + 1}\right) \tag{26}$$

According to Eqs. (23) and (26), the average droplet size depends on k_{eff} as well as on $k_c H_0$ and in **Figure 4** we have plotted $\langle R \rangle$ against k_{eff} for a certain value of $k_c H_0$ corresponding to a gemini surfactant that forms elongated micelles in water [13]. It is seen that the more simple equation (26) is an excellent approximation in virtually the entire regime of k_{eff} values. As expected, $\langle R \rangle$ is found to increase considerably with increasing values of $\bar{k}_c$ as well as decreasing values of H_0. The dependence on the more shape-related bending rigidity is more complex. For the case where $k_c \gg \bar{k}_c + k_s$, Eq. (26) simplifies to $\langle R \rangle = 1/H_0$ and droplet size only depends on spontaneous curvature. For small k_c values, however, the size of the droplets is explicitly determined by the quantity k_c times H_0.

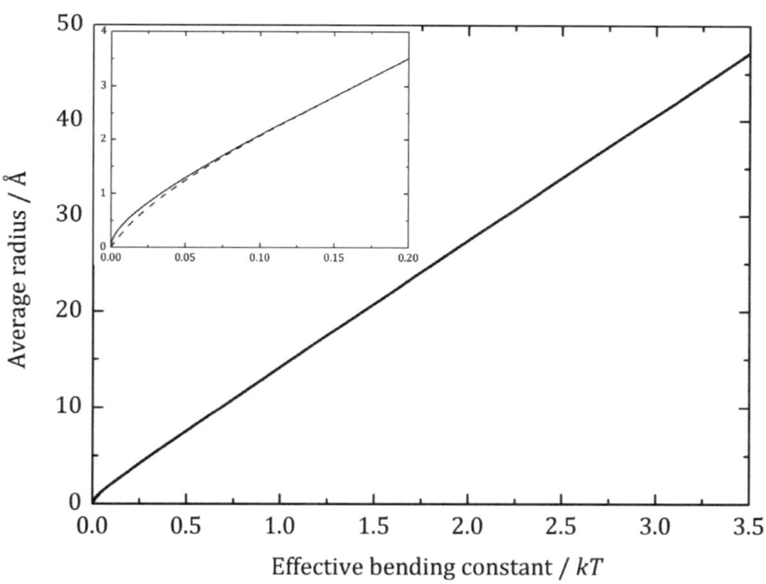

Figure 4. The average radius $\langle R \rangle$ of microemulsion droplets plotted against effective bending constant k_{eff} in accordance with Eqs. (23) and (24) (solid line) and Eq. (26) (dashed line). The spontaneous curvature H_0 was set to the same value as in **Figure 2**.

In addition to the three bending elasticity constants, the droplet size also depends on the self-assembly entropy parameter k_s. The always negative parameter k_s reduces the size of the droplets and its effect becomes stronger as the concentration of droplets decreases. As a result, microemulsion droplets grow in size with increasing surfactant concentration (cf. **Figure 2**). We may note that the expression $\langle R \rangle = (2k_c + \bar{k}_c)/2k_c H_0$, according to which $\langle R \rangle$ does not depend on surfactant concentration, was derived by Safran [24] for the case where self-assembly entropy effects were neglected. This is the optimal value of $\langle R \rangle$, obtained for the case where only bending elasticity properties are considered. As a matter of fact, the Safran equation is recovered as a special case of Eq. (26) in the limit $2k_c + \bar{k}_c \rightarrow \infty$, as k_s becomes negligible. Since k_s is always negative, the size of the droplets must be smaller than the value predicted by the Safran equation (cf. **Figure 2**).

6. Polydispersity

Like average droplet size, microemulsion droplet polydispersity may be predicted from the present model. From the size distribution function in Eq. (16), we may derive the following expression for the average of the squared radius

$$\langle R^2 \rangle = \frac{1}{\phi_{drop}} \int_0^\infty R^2 \phi(R) dR = \left(\frac{k_{eff}}{2k_c H_0} \right)^2 \frac{1}{y} \tag{27}$$

where

$$y = \frac{\sqrt{\pi} 2\alpha(2\alpha + 1)(1 + erf\sqrt{\alpha}) + 4\alpha^{3/2} e^{-\alpha}}{\sqrt{\pi}(4\alpha^2 + 12\alpha + 3)(1 + erf\sqrt{\alpha}) + 2\alpha^{1/2}(2\alpha + 5)e^{-\alpha}} \tag{28}$$

Hence, the polydispersity in terms of the relative standard deviation equals

$$\frac{\sigma_R}{\langle R \rangle} = \sqrt{\frac{\langle R^2 \rangle}{\langle R \rangle^2} - 1} = \sqrt{\frac{x^2}{y} - 1} \tag{29}$$

in accordance with Eqs. (23) and (27) and where the definition of x is given in Eq. (24). Equation (29) may be simplified so as to give the following excellent approximation for the polydispersity as a function of $\alpha = 4\pi k_{eff}/kT$ (cf. **Figure 5**).

$$\frac{\sigma_R}{\langle R \rangle} = \sqrt{\frac{4\alpha^2 + 3}{2\alpha(2\alpha + 3)^2}} = \sqrt{\frac{1}{6\alpha} + \frac{2}{6\alpha + 9} - \frac{4}{(2\alpha + 3)^2}} \tag{30}$$

The polydispersity according to Eqs. (29) and (30), respectively, is plotted in **Figure 5**. Notably, the droplet polydispersity does not at all depend on spontaneous curvature, but is seen to be a sole function of $k_{eff} = 2k_c + \bar{k}_c + k_s$. This means that k_{eff} may be directly determined from experimentally available polydispersities. It is seen that $\sigma_R/\langle R \rangle$ decreases with increasing values of the effective bending constant and it approaches $\sigma_R/\langle R \rangle = \sqrt{kT/8\pi k_{eff}}$ in the limit of large k_{eff}. The latter expression (with $k_s = 0$) was first derived by Milner and Safran for the case where self-association entropy effects were neglected [7]. Since the entropy parameter k_s depends on surfactant concentration [cf. Eq. (22)], $\sigma_R/\langle R \rangle$ according to Eq. (30) is predicted to decrease with increasing surfactant concentration.

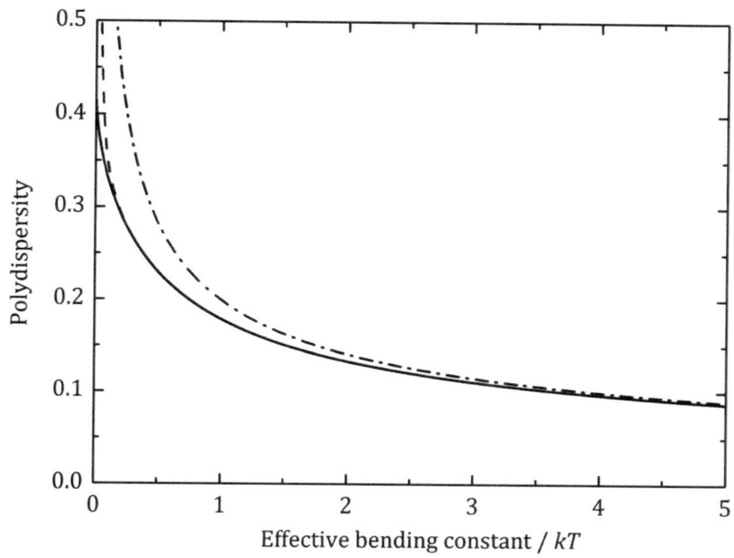

Figure 5. The relative standard deviation $\sigma_R/\langle R\rangle$ plotted against effective bending constant k_{eff} in accordance with Eq. (29) (solid line) and Eq. (30) (dashed line) as well as according to the approximate expression $\sigma_R/\langle R\rangle = \sqrt{kT/8\pi k_{\text{eff}}}$ (dash-dotted line).

7. Interfacial tension

The (planar) interfacial tension between the water and oil phases in a microemulsion in equilibrium with excess solvent phase is an experimentally available quantity. Hence, it has been observed that unusually small interfacial tensions are, in general, generated in microemulsion systems. The values are found to decrease in magnitude as the droplet size increases and may approach as low values as 10^{-3}–10^{-2} mNm^{-1} in non-ionic surfactant microemulsion systems at the phase inversion temperature (PIT) [10, 25].

From our present model we obtain the planar interfacial tension γ_p from the definition of k_{eff} in Eq. (19), i.e.

$$\gamma_p = \frac{4(k_c H_0)^2}{k_{eff}} \tag{31}$$

Similar to droplet size, the interfacial tension depends on the three bending elasticity constants as well as the entropy parameter k_s. In **Figure 6** we have plotted γ_p against the effective bending constant k_{eff} for some different values of the spontaneous curvature H_0. It is seen that γ_p decreases significantly as k_{eff} is increased or H_0 is reduced. For spontaneous curvatures approaching the limit $H_0 = 1/4\xi$, where bilayer structures start to form in aqueous solvents [26], γ_p may drop well below 1 mNm^{-1}. Notably, γ_p decreases with increasing surfactant concentration and droplet volume fraction since it, according to Eq. (31), depends on the entropy parameter k_s.

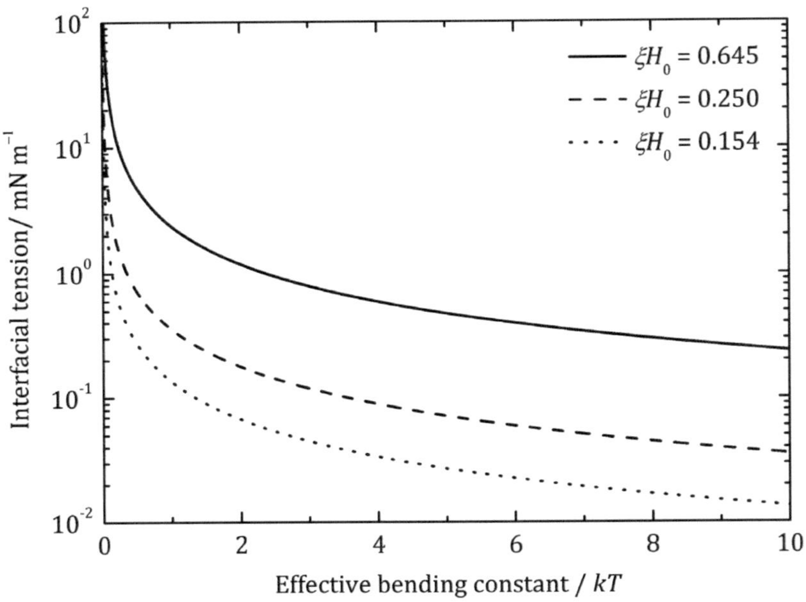

Figure 6. The planar interfacial tension γ_p of microemulsions plotted against effective bending constant k_{eff} in accordance with Eq. (31) for some different values of the spontaneous curvature H_0.

By means of combining Eqs. (26) and (31), we may eliminate $k_c H_0$ to deduce the following expression that relates interfacial tension and droplet size

$$\gamma_p = \frac{k_{eff}}{\langle R \rangle^2}\left(1 + \frac{2}{8\pi k_{eff}/kT + 1}\right)^2 \tag{32}$$

According to Eq. (32), the interfacial tension must decrease in magnitude as the size of the droplets increases (cf. **Figure 7**). This behaviour agrees very well with experimental observations that surfactants forming larger microemulsion droplets are, in general, found to have smaller interfacial tensions. γ_p is seen to drop below 1 mNm^{-1} as the radius exceeds a few nanometres for some typical value of $2k_c + \bar{k}_c$. Moreover, the interfacial tension is seen to fall below about 0.01 mNm^{-1} as the droplet size approaches orders of magnitudes equal to 1 μm. These values agree very well with observations of minimum interfacial tensions obtained for microemulsions formed by temperature sensitive non-ionic surfactants at PIT, where the microemulsion contains equal amounts of oil and water [10].

Since k_{eff} is directly related to the droplet polydispersity, it may, in principle, be eliminated by means of combining Eqs. (30) and (32) to yield a relation between the three experimentally available quantities $\langle R \rangle$, $\sigma_R/\langle R \rangle$ and γ_p. The following approximate expression (valid for polydispersities lower than about $\sigma_R/\langle R \rangle = 0.1$) is obtained by means of inserting $\sigma_R/\langle R \rangle = \sqrt{kT/8\pi k_{eff}}$ in Eq. (32)

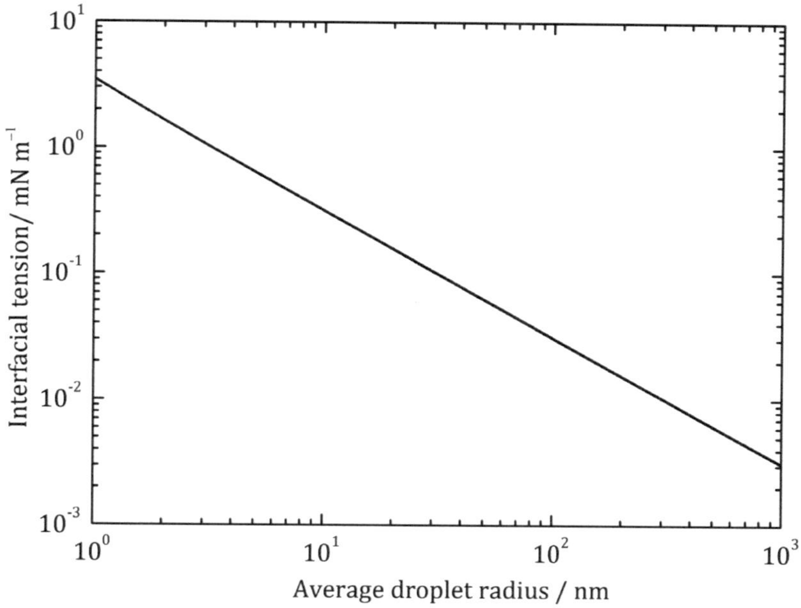

Figure 7. The planar interfacial tension γ_p of microemulsions plotted against the average droplet radius $\langle R\rangle$ in accordance with Eq. (32).

$$\gamma_p = \frac{kT}{8\pi\langle R\rangle^2(\sigma_R/\langle R\rangle)^2}\left(1+\frac{2}{(\sigma_R/\langle R\rangle)^{-2}+1}\right)^2 \quad (33)$$

The relation between interfacial tension, droplet radius and polydispersity may also be expressed in terms of the dimensionless parameter [27]

$$w \equiv 8\pi\left(\frac{\sigma_R}{\langle R\rangle}\right)^2\frac{\gamma_p\langle R\rangle^2}{kT} \quad (34)$$

Combining Eqs. (26), (30) and (31), the following approximate expression for w may be derived

$$w = \frac{4\alpha^2+3}{(2\alpha+1)^2} = 1-\frac{2(8\pi k_{eff}-1)}{(8\pi k_{eff}+1)^2} \quad (35)$$

In the limit $k_{\text{eff}}\rightarrow\infty$ (equivalent to $\sigma_R/\langle R\rangle\rightarrow 0$), Eq. (35) may be further simplified so as to give $w = 1$, the expression of which has been derived by Gradzielski et al. [27] for the case where self-association entropy effects were neglected. It is necessary that γ_p assumes low values for a system of small droplets, with a large overall interfacial area, to form. The ultra-low values of interfacial tensions experimentally observed in microemulsion systems appear as a consequence of the comparatively large interfacial area between oil and water that is formed in colloidal systems consisting of small droplets. Microemulsions are thermodynamically equilibrated systems and the disadvantageous (positive) free energy of increasing the interfacial area is exactly balanced by the favourable (negative) free energy of mixing oil droplets with solvent water. The entropy of mixing increases with the number of dissolved droplets, equivalent to a reduction in droplet size, and the corresponding contribution to the free energy becomes increasingly more negative with decreasing droplet size. As a result, the interfacial tension must decrease in magnitude with increasing droplet size in accordance with Eq. (33).

8. Solubilisation

A notable property of microemulsion systems is the ability to dissolve hydrophobic components which otherwise are insoluble in an aqueous solvent. A quantitative measure of the solubilisation capacity may be considered as the ratio of the volume of oil (V_{oil}) that is solubilized and the volume of surfactant (V_{surf}) present in the solution. Hence, we may define the solubilisation capacity as follows

$$\sigma \equiv \frac{V_{\text{oil}}}{V_{\text{surf}}} \tag{36}$$

Since the surfactant is exclusively located at the droplet interface whereas the oil molecules are confined to the interior of the droplets, σ becomes strongly dependent on droplet size. The volumes of solubilized oil and aggregated surfactant, respectively, are obtained from the following geometrical relations valid for spherical droplets with an average radius $\langle R \rangle$ and an interfacial surfactant monolayer with thickness ξ (*cf.* **Figure 1**)

$$V_{\text{oil}} = N_{\text{drop}} \cdot \frac{4}{3}\pi(\langle R \rangle - \xi)^3 \tag{37}$$

$$V_{\text{surf}} = N_{\text{drop}} \cdot \frac{4}{3}\pi\langle R \rangle^3 - V_{\text{oil}} = N_{\text{drop}} \cdot \frac{4}{3}\pi[\langle R \rangle^3 - (\langle R \rangle - \xi)^3] \tag{38}$$

where N_{drop} is the total number of droplets. Combining Eqs. (36)–(38) gives the following expression for the solubilisation capacity

$$\sigma = \frac{1 - \delta^3 + 3\delta^2 - 3\delta}{\delta^3 - 3\delta^2 + 3\delta} \tag{39}$$

as a function of the single quantity δ defined as

$$\delta \equiv \frac{\xi}{\langle R \rangle} = \frac{2k_c \xi H_0}{k_{\text{eff}}}\left(1 - \frac{2}{8\pi k_{\text{eff}}/kT + 3}\right) \tag{40}$$

The second equality in Eq. (40) is obtained by means of employing the expression in Eq. (26) for the average droplet radius.

In **Figure 8**, we have plotted the average droplet radius $\langle R \rangle$ against the volume fraction of surfactant self-assembled in the droplet interfaces. It is seen that the droplet size increases with an increasing surfactant concentration as a result of the self-association entropy effects

http://dx.doi.org/10.5772/67369

discussed above. This means that we obtain an enhanced solubilisation effect as the amount of added surfactant is increased. The number of droplets, as well as the average size of the droplets, increases as the surfactant concentration is raised and both effects contributes to an increased solubilisation capacity.

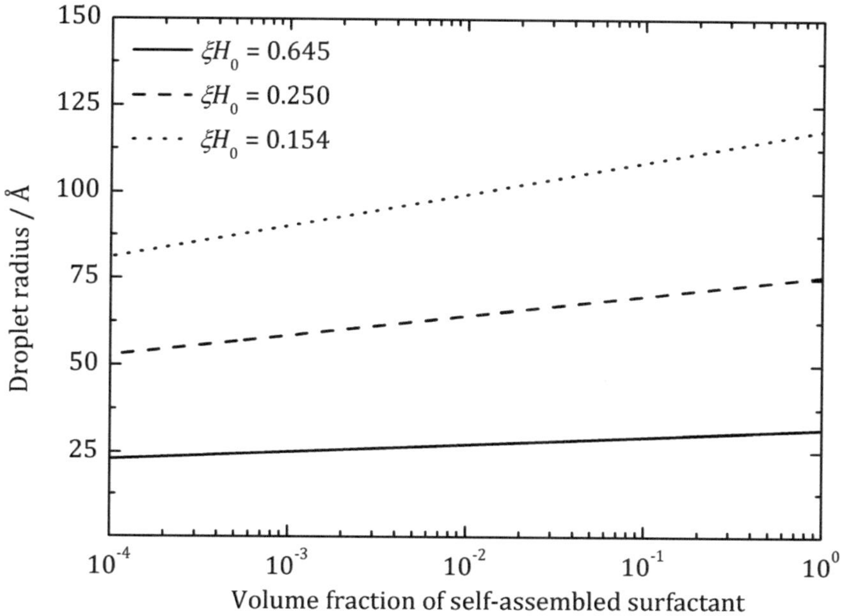

Figure 8. The average radius $\langle R \rangle$ of microemulsion droplets plotted against the volume fraction φ_s of self-assembled surfactant in accordance with Eqs. (23) and (24) for some different values of the spontaneous curvature H_0. The bending rigidity k_c and saddle-splay constant $\bar{k}_c$ assume identical values as in **Figure 2**.

The solubilisation capacity σ, as defined in Eq. (39), is plotted against the surfactant concentration in **Figure 9**. It is seen that σ increases considerably with increasing surfactant concentration. σ strongly depends on the different bending elasticity constants. The bending rigidity k_c and saddle-splay constant $\bar{k}_c$ have been found to be of the order of magnitude kT for many common surfactants and the possibility to tune the values of them is usually limited [11]. The spontaneous curvature H_0, on the other hand, strongly depends on the surfactant HLB value and may assume values from close to zero, for amphiphilic molecules forming bilayers, to much higher values for surfactants forming small micelles. In **Figure 9** we have chosen to compare σ for three different values of H_0. The smallest value $\xi H_0 = 1.55^{-1} = 0.154 > 0.25$ corresponds to a (gemini) surfactant that form elongated micelles in an aqueous solution and the solubilisation capacity is found to be rather small, with σ considerably lower than unity. However, σ is found to increase substantially as the spontaneous curvature is decreased to $\xi H_0 = 0.25$, which corresponds to the point of transition from micelles to bilayers in an aqueous solvent in the absence of oil [26]. Further decreasing H_0 to $\xi H_0 = 6.5^{-1} = 0.645$, i.e. to values where rather large bilayer structures are expected to form, the solubilisation capacity σ exceeds unity, which means that the amount of oil that is dissolved exceeds the amount of surfactant present in the solution. As a matter of fact, these results agree very well with observations that the solubilisation capacity of decane (oil) in mixed tetradecyldimethylamine oxide (TDMAO)-hexanol microemulsion droplets was found to be significantly larger at high hexanol compositions, where vesicles are formed in the absence of oil, as compared to compositions where mixed micelles form in the absence of oil [5].

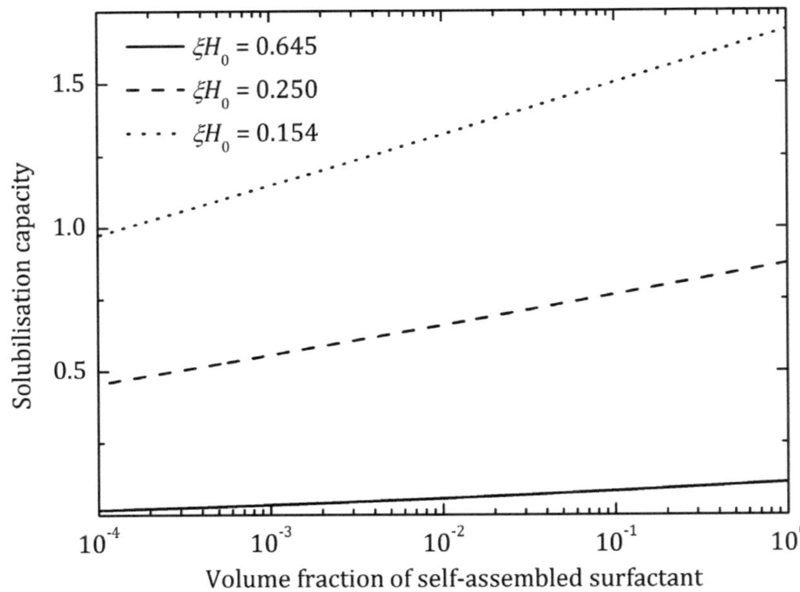

Figure 9. The solubilisation capacity σ of a microemulsion plotted against the volume fraction ϕ_s of self-assembled surfactant in accordance with Eq. (39) for some different values of the spontaneous curvature H_0. The bending rigidity k_c and saddle-splay constant $\bar{k}_c$ assume identical values as in **figure 2**.

9. Summary

A comprehensive theoretical model for the formation of microemulsion droplets has been presented. The theory is based on thermodynamics of self-assembling surfactant molecules aggregated in the droplet interfaces, combined with bending elasticity properties of the surfactant monolayer as taken into account by the three parameters spontaneous curvature, bending rigidity and saddle-splay constant. It relates properties depending on surfactant chemical structure with experimentally available properties of the droplets, such as average size, polydispersity, interfacial tension and solubilisation capacity. It has recently been demonstrated that all three constants H_0, k_c and $\bar{k}_c$ may be determined from measurements of growth behaviour of surfactant micelles in aqueous solutions (in the absence of oil). [12, 13, 22] Hence, from the present model it is possible to predict, for instance, the solubilisation capacity of oil of a certain surfactant by studying the growth behaviour of micelles that is formed by the surfactant in the absence of oil.

Likewise, it is possible to determine the various bending elasticity constants by means of measuring the average droplet radius, polydispersity and/or interfacial tension of a microemulsion system. We have included the important contribution of entropy of self-assembly and, as a result, we obtain quantitative expressions that differ from previously derived ones [7, 24, 27] where self-association entropy effects were omitted. Nevertheless, the previously derived expressions for average droplet radius, polydispersity and interfacial tension are all recovered as special cases from our model in the limit $2k_c + \bar{k}_c \rightarrow \infty$. Taking into account self-association entropy effects, quantified in terms of the parameter k_s, have the impact on the model to predict smaller values of the droplet size as compared to the case where the entropy of self-assembly is neglected. Moreover, k_s explicitly depends on surfactant concentration (k_s increases with increasing concentration), implying that the droplet size is predicted to increase, whereas the polydispersity decreases, with increasing surfactant concentration. Moreover, the model predicts ultra-low values of the interfacial tension that agrees excellently with experimental measurements [10, 25]. Likewise, the capacity to solubilize a hydrophobic component (oil) is predicted to increase dramatically by means of reducing the spontaneous curvature (equivalent to decreasing the HLB value) of the surfactant, in agreement with experiments. In addition, self-association entropy effects are shown to generate an enhanced solubilisation capacity as the size of the microemulsion droplets is found to increase with increasing surfactant concentration.

Author details

L. Magnus Bergström

Address all correspondence to: magnus.bergstrom@farmaci.uu.se

Department of Pharmacy, Uppsala University, Sweden

References

[1] Winsor PA. Binary and multicomponent solutions of amphiphilic compounds solubilization and the formation, structure and theoretical significance of liquid crystalline solutions. Chem Rev. 1968;68:1–40.

[2] Jahn W, Strey R. Microstructure of microemuisions by freeze fracture electron microscopy. J Phys Chem. 1988;92:2294–301.

[3] Griffin WC. Classification of surface-active agents by "HLB". J Soc Cosmet Chem. 1949;1:311–26.

[4] Kuneida H, Shinoda K. Phase behavior in systems of nonionic surfactant/water/oil around the hydrophile-lipophile-balance-temperature (HLB-temperature). J Dispers Sci Technol. 1982;3:233–44.

[5] Gradzielski M, Hoffmann H, Langevin D. Solubilization of decane into the ternary system TDMAO/1-Hexanol/Water. J Phys Chem. 1995;99:12612–23.

[6] Bergström LM. Influence of bending energetics on the size, shape and polydispersity of droplet microemulsions. Colloids Surf A. 2008;316:15–26.

[7] Milner ST, Safran SA. Dynamical fluctuations of droplet microemulsions and vesicles. Phys Rev A. 1987;36:4371–9.

[8] Eriksson JC, Ljunggren S, Henriksson U. A novel approach to the mechanics and thermodynamics of spherical micelles. J Chem Soc, Faraday Trans 2. 1985;81:833–68.

[9] Helfrich W. Elastic properties of lipid bilayers: Theory and possible experiments. Z Naturforsch C. 1973;28:693–703.

[10] Sottmann T, Strey R. Ultralow interfacial tensions in water–n-alkane–surfactant systems. J Chem Phys. 1997;106:8606–15.

[11] Gradzielski M. Bending constants of surfactant layers. Curr Opinion Colloid Interface Sci. 1998;3:478–84.

[12] Bergström LM. Explaining the growth behavior of surfactant micelles. J Colloid Interface Sci. 2015;440:109–18.

[13] Bergström LM, Tehrani-Bagha AR, Nagy G. Growth behavior, geometrical shape, and second CMC of micelles formed by cationic gemini esterquat surfactants. Langmuir. 2015;31:4644−53.

[14] Bergström LM. Bending elasticity of charged surfactant layers: The effect of layer thickness. Langmuir. 2006;22:3678–91.

[15] Bergström LM. Bending elasticity of charged surfactant layers: The effect of mixing. Langmuir. 2006;22:6796–813.

[16] Bergström LM. Bending elasticity of nonionic surfactant layers. Langmuir. 2009;25: 1949–60.

[17] Porte G. Lamellar phases and disordered phases of fluid bilayer membranes. J Phys: Condens Matter. 1992;4:8649–70.

[18] Safran SA. Curvature elasticity of thin films. Adv Phys. 1999;48:395–448.

[19] Kozlov MM, Helfrich W. Effects of a cosurfactant on the stretching and bending elasticities of a surfactant monolayer. Langmuir. 1992;8:2792–7.

[20] Porte G, Ligoure C. Mixed amphiphilic bilayers: Bending elasticity and formation of vesicles. J Chem Phys. 1995;102:4290–8.

[21] Bergström LM, Garamus VM. Geometrical shape of micelles formed by cationic dimeric surfactants determined with small-angle neutron scattering. Langmuir. 2012;28:9311–21.

[22] Bergström LM, Grillo I. Correlation between geometrical shape and growth behaviour of surfactant micelles investigated with small-angle neutron scattering. Soft Matter. 2014;10:9362–72.

[23] Eriksson JC, Ljunggren S. Thermodynamic evaluation of the polydispersity of droplet microemulsions. Langmuir. 1995;11:1145–53.

[24] Safran SA. Saddle-splay modulus and the stability of spherical microemulsions. Phys Rev A. 1991;43:2903–4.

[25] Sottmann T, Strey R. Evidence of corresponding states in ternary microemulsions of water-alkane-CiEJ. J Phys: Condens Matter. 1996;8:A39–A48.

[26] Bergström LM. Model calculations of the spontaneous curvature, mean and Gaussian bending constants for a thermodynamically open surfactant film. J Colloid Interface Sci. 2006;293:181–93.

[27] Gradzielski M, Langevin D, Farago B. Experimental investigation of the structure of nonionic microemulsions and their relation to the bending elasticity of the amphiphilic film. Phys Rev E. 1996;53:3900–19.

Chapter 7

Self-Propelled Motion of Micrometer-Sized Oil Droplets in Aqueous Solution of Surfactant

Taisuke Banno, Taro Toyota and Kouichi Asakura

Additional information is available at the end of the chapter

http://dx.doi.org/10.5772/67249

Abstract

When an immiscible oil is dispersed in an aqueous solution of a surfactant, emulsions consisting of various-sized oil droplets are generated. Micrometer-sized oil droplets exhibit exotic dynamics such as self-propelled motion in the surfactant solution. Transfer of the surfactant from the aqueous solution phase to the oil droplets through their interface leads to the self-propelled motion in a far-from-equilibrium condition. In this chapter, we demonstrate the observation methods of the self-propelled motion of micrometer-sized oil droplets using phase-contrast, polarized, and fluorescence microscopes and discuss their motion mechanism. Since the generated self-assemblies in micrometer-sized droplet systems are difficult to be identified by spectroscopic methods, the mechanisms of their self-propelled motion have not been clarified. When they are fully understood from nano- to microscale, these findings may be useful to develop not only more stable emulsion systems but also droplet-type analysis systems at the micrometer scale that can carry out reaction, analysis, and detection automatically without the need for an external force.

Keywords: cationic surfactant, chemical reaction, far-from-equilibrium state, interfacial tension, Marangoni effect, micrometer-sized oil droplet, optical microscope, self-propelled motion

1. Introduction

Oil and water are immiscible liquids. When oil is added to water and the mixture is stirred vigorously, oil droplets and water droplets are dispersed in the oil and water phase, respectively. This increases the total area at the oil-water interface; therefore, such an emulsion is unstable. As a result, droplets aggregate and fuse gradually to minimize the contact area between the oil and water phase. To improve the stability of emulsions and utilize those emulsions as functional capsules in fields of cosmetics, pharmaceuticals, and foods, many physical and chemical approaches have been developed. Therefore, emulsion science and technology is necessary for an improvement in quality of life. In this chapter, we do not describe a stabilization technology for emulsions but rather a research trend in the behavior of micrometer-sized droplets within emulsions in a far-from-equilibrium state. Namely, the phenomena of oil droplets moving three-dimensionally (self-propelled motion) in a ternary system composed of water, oil, and surfactant and the methods to measure this motion are described.

Surfactant molecules form various types of self-assembly in water or buffered aqueous solution [1]. At a relatively low surfactant concentration in water, colloidal self-assemblies, such as spherical, disklike, rodlike, or wormlike micelles, are formed spontaneously. When a small amount of the oil components is added to such a colloidal system, it is solubilized within micelles. An electron microscope is required for the observation of these swelling micelles (microemulsions). They are stable thermodynamically, and this is defined as an equilibrium condition in this chapter. When the oil component is further increased, emulsions consisting of oil droplets with diameters ranging from nanometer to submillimeter are formed. For example, upon the addition of an oil component that was almost insoluble in water, such as *n*-hexane and chloroform, to aqueous surfactant solution of a relatively low concentration, the dispersion became turbid, and, in a short while, it separated into water and oil phase. However, when some specific oil components were dispersed into a similar surfactant solution, the dispersion was not separated into two phases but turned turbid over a week. Interestingly, immediately after the addition of the oil component, it was observed under an optical microscope that oil droplets with diameters ranging from 5 to 100 μm were self-propelled in a dispersion that was partly turbid (**Figure 1**). Self-propelled motion of these spherical droplets had a locomotion speed of above 5 μm/s, which was significantly faster than the random walk of micrometer-sized colloidal particles. In other words, the observed self-propelled motion could be distinguished from Brownian motion. Though the self-propelled motion of these droplets ceased within an hour, the spherical droplets were stable. This phenomenon was observed in a far-from-equilibrium state that gradually approached equilibrium eventually and is of great interest because the micrometer-sized objects moved without the addition of external forces. Note that these observed results depended on the intensity of stirring and composition of emulsion system.

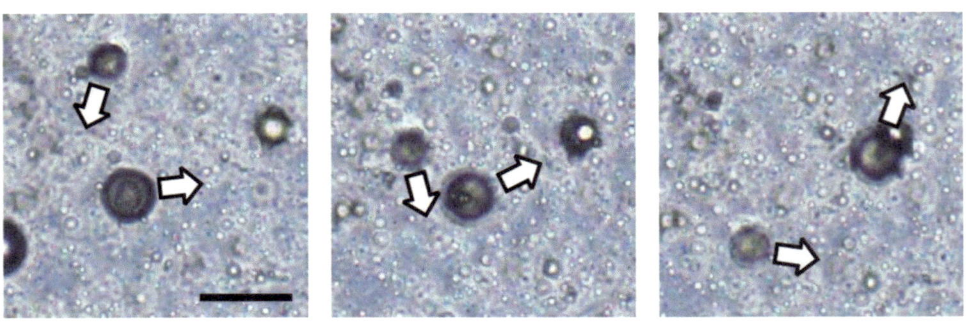

Figure 1. Typical sequential micrographs (time interval = 3 s) of self-propelled motion by micrometer-sized spherical oil droplets in a ternary system composed of water, oil (*n*-heptyloxybenzaldehyde), and surfactant (*N*-hexadecyl-*N,N,N*-trimethylammonium bromide) at room temperature (23–25°C). The arrows indicate the direction of the self-propelled oil droplets. Bar: 100 μm.

2. Observations of micrometer-sized self-propelled oil droplets

2.1. Optical microscopes for the observations of oil droplets in aqueous solution

A bright-field stereomicroscope can be used to observe micrometer-sized oil droplets in aqueous solutions when oil droplets are thick and their refractive index is significantly different from their surrounding medium (the bulk solution). However, when the difference in refractive index is small and the oil droplets are thin, specific microscopes are required for observation. In this section, the operating principles of polarized, phase-contrast, and fluorescence microscopes, which are used regularly for the observation of soft matter, including oil droplets at the micrometer scale, are introduced briefly.

2.1.1. *Polarized microscope*

When the surfactant concentration is relatively high in an emulsion system, the nematic and lamellar structures of lyotropic liquid crystal phases are formed. Since the interaction of each phase with light differs because of the different molecular orientation, the structures can be characterized by the texture of polarized microscopy images. In the system of a polarized microscope, the transmitted light through samples, placed on a stage between two polarizers oriented at 90° to the illumination, is observed. Since birefringence occurs because of the optical anisotropy of phase in the sample, the phases within the sample can be identified from the observed textures. For example, to investigate the stability of a ternary system, Abe et al. investigated dimyristoylphosphatidylcholine/water/saturated hydrocarbon, where propanol was added as a cosurfactant, and clarified the composition that generated a stable oil-in-water microemulsion using a polarized microscope [2]. In addition, Ho et al. observed a dispersion prepared by stirring lauryl or cetyl alcohol with a sodium lauryl sulfate aqueous solution using ultrasonication, under a polarized microscope [3]. By varying the alcohol concentration and temperature, various phases such as the schlieren textures of the nematic liquid crystal phase were confirmed. No textures were observed in self-propelled oil droplets under a polarized microscope unless water-insoluble molecules exhibiting a thermotropic liquid crystalline phase were used, indicating that they were not in a liquid crystal phase but an isotropic liquid phase. Self-propelled micrometer-sized droplets comprising thermotropic liquid crystal phases in a surfactant solution have recently gained substantial attention regarding the topological defect of the droplet in self-propelled motion [4].

2.1.2. *Phase-contrast microscope*

Phase-contrast microscopes are suitable for the observations of transparent specimens, in which their refractive index is similar to that of the surrounding medium (such as living cells and bacteria). Self-propelled motion of micrometer-sized oil droplets in aqueous surfactant solution has also been observed under a phase-contrast microscope. Visualization of transparent specimens with a low refractive index has been achieved by utilizing the diffraction and interference of light. **Figure 2** shows the optical path and operating principle of visualization in this microscope system. A phase-contrast microscope has a ring slit in the condenser, and the objective lens is equipped with a phase-shift plate (**Figure 2A**). The ring-shaped illuminating light that passes the ring slit in the condenser is focused on the phase-shift plate and is guided to the image plane through an objective lens uniformly. However, in the presence of a specimen between the condenser and objective lens, some of the illuminating light is diffracted by the specimen and separated into two diffraction order beams (the +1 and −1 order) and one remaining light beam, which is unaffected by the specimen (zeroth-order diffracted light), known as the background light. The two diffracted light beams change with the direction of travel and are therefore not focused on the phase-shift plate. On the other hand, the zeroth-order diffracted light beam goes straight ahead and passes the phase-shift plate. Thus, three light beams are focused on the image plane of the objective lens. The image contrast is strengthened by the following two factors: the generated interference between the diffracted and background light rays in the regions of the field of view that contain the specimen and the reduction in the amount of background light that reaches the image plane.

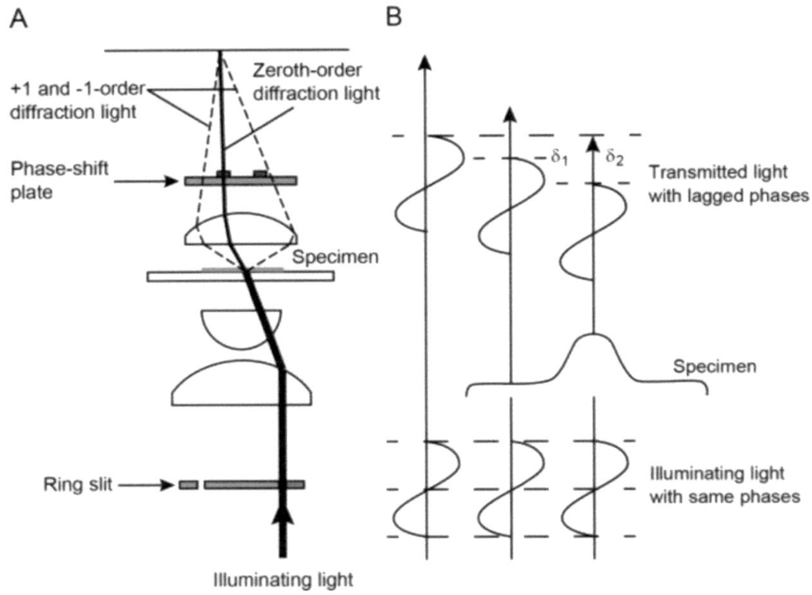

Figure 2. Schematic illustration of optics (A) and operating principle of visualization (B) of a phase-contrast microscope. δ_1 and δ_2 are the lagged phase between the transmitted light rays through the specimen and that of the background, respectively.

For example, if it is considered that the phase of the zeroth-order diffracted light is directed by the phase-shift plate, the light intensity is zero on the image plane where the three light beams interfere and, thus, the image of the specimen should have dark contrast in the bright field of view. In contrast, when the zeroth-order diffracted light is lagged compared to the other two diffracted light beams, the light intensity becomes maximum because of the interference between the three light beams. Therefore, the image of the specimen with bright contrast is observed in the dark field of view. On the basis of these principles, a transparent specimen, which is difficult to distinguish from the surrounding medium using a bright-field microscope, can be observed by strengthening image contrast.

On observing the self-propelled oil droplets using a phase-contrast microscope, it was found that the interior of the droplets had many small particles and they formed a convective flow. The direction linked with the inlet and outlet of this convective flow was the same as the self-propelling motion direction of the droplets, considering that their motion could be associated with the flow fields within the droplets.

2.1.3. *Fluorescence microscope*

Dynamics of molecules and particles which are labeled by fluorescent molecules and particles can be traced selectively. Therefore, fluorescence observations have been frequently used in the life sciences. A transparent specimen can be observed under a phase-contrast microscope. By using a fluorescence microscope simultaneously, the specific molecule and its distribution can be visualized in a specimen.

When a substance, which absorbs light energy and emits its energy as light, is illuminated by excitation light, such as X-rays, ultraviolet light, or visible light, the transition of its electrons from the ground state to the excited state occurs. However, because this excited state is unstable energetically, the electrons that absorb energy relax readily to their ground state. In this relaxation process, the emitted light is fluorescence. The relaxation time from an excited state to a ground state is short, below ~10 ms in general, and the luminescence is quenched due to fading of a fluorescent substance under a continuous excited light. Taking this into consideration, the system of a fluorescence microscope has a specific set of operation principles.

In general, if a fluorescent dye has a suitable molecular structure for emitting fluorescence in a specimen, the distribution and migration of a fluorescent dye can be detected easily. The distribution and migration of multiple structures are also observed simultaneously by using multiple fluorescent dyes. In addition, if the background brightness is lowered considerably, the detection sensitivity using a fluorescence microscope is much higher than when using other observation techniques. Therefore, this technique is frequently used for the observation of the dynamic behavior of self-assemblies formed by amphiphiles in aqueous solution. For example, micrometer-sized droplets in a ternary emulsion system composed of water, oil, and surfactant exhibited demulsification triggered by a photoisomerization of the photoreactive surfactant having an azobenzene group [5]. This phenomenon was clarified by fluorescence observations using a hydrophilic fluorescent dye, calcein, and a hydrophobic fluorescent dye, pyrene. In oil droplet systems that exhibit self-propelled motion, the fluorescence microscopy technique is considered to be effective for visualizing the convective flows of the droplets.

2.2. Mechanism of self-propelled motion of oil droplets

2.2.1. *Observation of self-propelled motion of oil droplets*

Self-propelled motion of micrometer-sized oil droplets has been observed in dispersions composed of specific oils and surfactants as shown in **Figure 3**. For example, in a dispersion prepared by adding a benzaldehyde-type oil component (**1**) to the aqueous solution of a cationic surfactant having a primary amine group at the end of its molecular structure (**2**), self-propelled motion of oil droplets accompanied by an internal convective flow was observed under a phase-contrast microscope. The micrometer-sized structures of closed membranes (giant vesicles) were stripped off at the inlet position of the convection at the surface of this droplet, releasing giant vesicles at the opposite side of self-propelled droplet. In addition, by using a fluorophore with molecular structure similar to **1**, having 4,4-difluoro-4-bora-3a,4a-diaza-s-indacene (BODIPY) group, it was confirmed by fluorescence observations that giant vesicles were formed by the production of an amphiphile (**3**) in the droplet. Therefore, the unidirectional motion of self-propelled droplets may be related to not only inner convection but also the evolution of giant vesicles at the opposite side of the droplets [6].

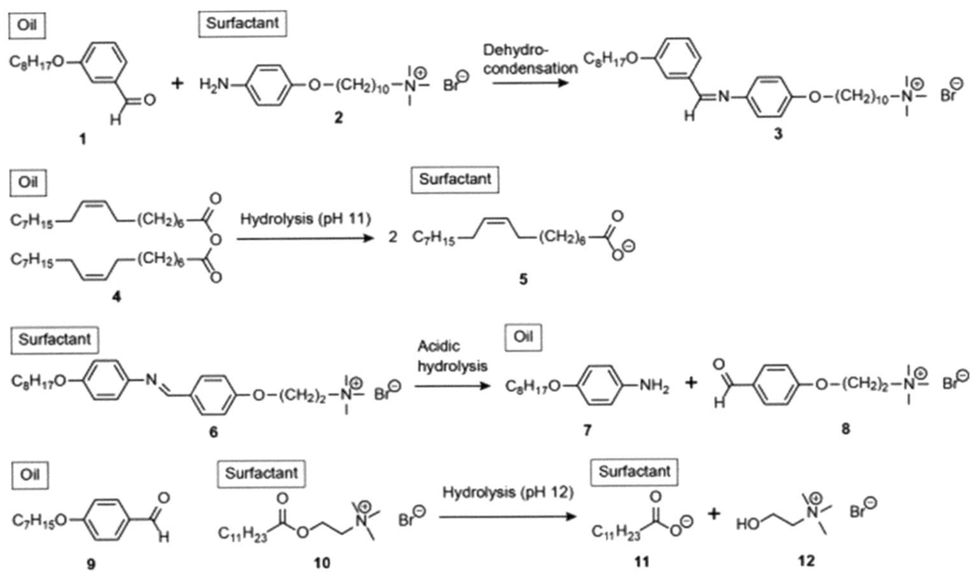

Figure 3. Reaction formula in emulsion systems in which micrometer-sized oil droplets were observed.

It has been found that self-propelled motion of oil droplets is caused by generated flow fields induced by the adsorption of surfactant molecules onto the droplet surface in dispersion. Hanczyc et al. reported that oleic anhydride (**4**) droplets containing nitrobenzene exhibited self-propelled motion in a solution composed of sodium hydroxide and sodium oleate (**5**) [7]. In addition, the droplets were self-propelled from a region of lower pH to that of higher pH in a certain gradient field. Thus, it was considered that the surface-active oleate, which was generated by basic hydrolysis of oleic anhydride, got adsorbed to the surface of the droplets, inducing the formation of flow fields. The unidirectional motion of droplets was a result of the flow fields caused by such local adsorption of surfactant molecules. Moreover, in a solution of an imine-containing cationic surfactant (**6**), aniline-derivative (**7**) droplets with diameters of 10–140 μm exhibited self-propelled motion accompanied by the evolution of waste oil droplets with diameters of 1–3 μm on the posterior surface. By tracing the behavior of the droplets using a fluorescence microscope, it was visualized that their dynamics were triggered by the generation of flow at the droplet surface accompanied by the hydrolysis of **6** to produce oil component **7** and electrolyte **8** [8]. Recently, it was also observed that oil droplets composed of *n*-heptyloxybenzaldehyde (**9**) moved autonomously as aggregates formed on their surface in an ester-containing cationic surfactant

http://dx.doi.org/10.5772/67249

(**10**) solution under a basic condition [9]. This was likely because of the production of lauric acid (**11**) and electrolyte **12** by the hydrolysis of surfactant **10** and the subsequent neutralization of the fatty acid with sodium hydroxide, forming a cation-anion complex surfactant because of the ionic bonding. These results indicate that the molecular conversion of oil or surfactant component is a key factor, which influences the driving force of droplet motion. In addition, oil droplets composed of benzaldehyde-derivative **9** were self-propelled in a solution of nonreactive surfactants, such as *n*-alkyltrimethylammonium bromide (CnTAB, n = 12, 14, 16) or gemini-type cationic surfactants having two hydrophobic and hydrophilic groups [10]. Thus, it is considered that self-propelled motion of oil droplets is strongly associated with not only the molecular conversion but also the molecular interactions between the surfactant and oil molecules.

2.2.2. Proposed mechanism for self-propelled motion of oil droplets

On the basis of the above findings, the mechanism of self-propelled motion of oil droplets is interpreted as follows (**Figure 4**). In a relatively concentrated surfactant solution, self-assembly of surfactant molecules, such as spherical and disklike micelles, distributes heterogeneously [11, 12]. On adding of an oil component into such a surfactant solution, oil droplets with various submillimeter sizes are formed. Surfactant molecules begin to adsorb onto the droplet surfaces, and heterogeneity of droplet surfaces is induced by nonuniform surfactant concentrations, as well as by thermal fluctuations. This causes an imbalance in the interfacial tension between sites with adsorbed surfactant molecules (lower interfacial tension) and bare sites (higher interfacial tension) on the droplet surface. The flow at the oil droplet surface based on the imbalance of the interfacial tension is maintained by Marangoni instability, indicating symmetry breaking [13]. Marangoni flow and subsequent mass transfer are likely caused by relatively strong intermolecular interactions between the surfactant and oil molecules. Furthermore, the momentum between inside and outside of the droplet is exchanged through the Marangoni flow, and the droplet itself is driven in a certain direction. These processes can be considered with regard to interfacial energy: the interfacial energy of the droplet's leading edge (where surfactants are adsorbed) is smaller than that of its trailing edge, inducing a slight movement of droplets including dynamic interfacial fluctuation. The more surfactant the moving droplet takes on, the more the droplet continues to move because of the flux balance between the oil droplet and the bulk solution. At the foreside of self-propelled motion, larger amounts of surfactant molecules adsorb to the droplet surface, and self-propelled motion occurs because of the feedback mechanism caused by the sustainment of flow fields. Thus, self-propelled motion could be affected by both the attractive interactions between the oil and surfactant molecules and the mobility of surfactant molecules at the droplet surface.

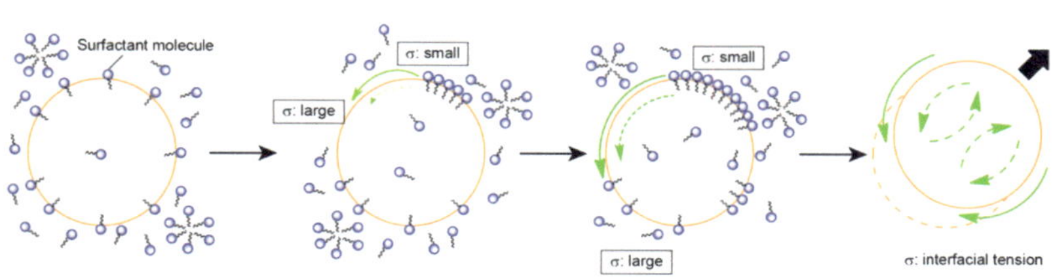

Figure 4. Schematic representation of the proposed mechanism for self-propelled motion of oil droplets. Surfactant molecules are omitted in the fourth image. σ, local interfacial tension of oil droplet surface.

2.2.3. *Visualization of the surrounding and internal flow fields of self-propelled oil droplets*

To verify the proposed mechanism of self-propelled motion of oil droplets experimentally, the surrounding and internal flow fields of droplets that were observed under a phase-contrast microscope should be analyzed in detail. These flow fields have been visualized, and their motion speed was analyzed using particle image velocimetry (PIV). This is a method that analyzes velocity and its vector of distinguishable particles (tracers), such as micrometer-sized fluorescent beads, in the flow fields by the following procedure. Firstly, the movement of dispersed tracers in the flow fields is monitored successively at a certain time interval. Secondly, the specified region containing some tracers in the nth image is correlated with a region in the $[n + 1]$th image. From the comparison of these two images, the velocity and direction of tracers are calculated by the detection of the moving distances between them (**Figure 5**).

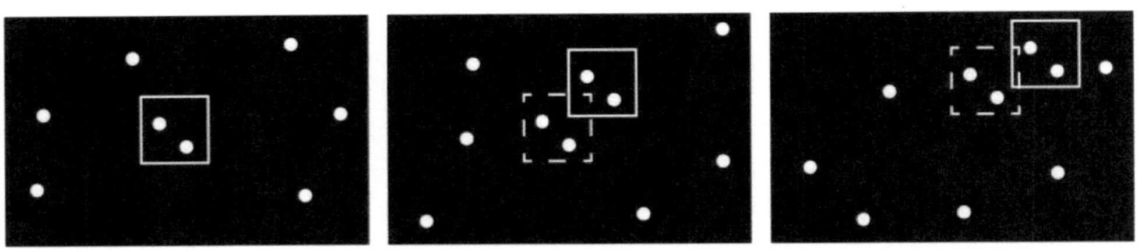

Figure 5. Principle of particle image velocimetry (PIV) measurement. The velocity vector of tracers (white circle) in the square region is calculated by the moving distance for a certain time interval (Δt).

Figure 6A shows the PIV results of self-propelled motion of oil droplets composed of benzaldehyde-derivative (**9**) in 50 mM C16TAB solution using micrometer-sized fluorescent polystyrene beads as a tracer. Hydrophilic fluorescent beads were used for measurements of the surrounding flow fields of the droplets. The surrounding flow fields observed in a pseudo-two-dimensional manner under a fluorescence microscope are shown in **Figure 6A**. The velocity of droplets was constant at ca. 20 μm/s. The velocity of the flow fields was accelerated and became maximum when tracers moved in the direction of a self-propelled droplet, which was from the region a to b along the droplet surface. On the other hand, it decreased when tracers moved in the opposite direction from region b to c. These results suggest that Marangoni flow occurred from the leading edge to trailing edge at the surface of the self-propelled droplets.

In contrast, hydrophobic fluorescent beads were dispersed into benzaldehyde derivative **9** for observations of the internal flow fields of the droplets. As a result, some flow fields were confirmed as indicated by white arrows in **Figure 6B**. By tracing the fluorescent bead carefully, it was found that the velocity of flow accelerated at the center (b′) and the forward position (c′) of the self-propelled droplet. The strong flow at the forward position may be derived from the flow of molecular assemblies composed of mainly oil components near the droplet surface accompanied with Marangoni flow. In addition, flow at the center may be because of molecular assemblies flowing from the backward to the forward position. Even though the PIV technique cannot produce precise information on tracers in terms of time and space resolution in

http://dx.doi.org/10.5772/67249

three-dimensional space, the proposed mechanism was supported by the pseudo-two-dimensional analyses discussed. Therefore, the analysis of flow fields using PIV is highly effective in the investigation of the mechanism of self-propelled motion of oil droplets.

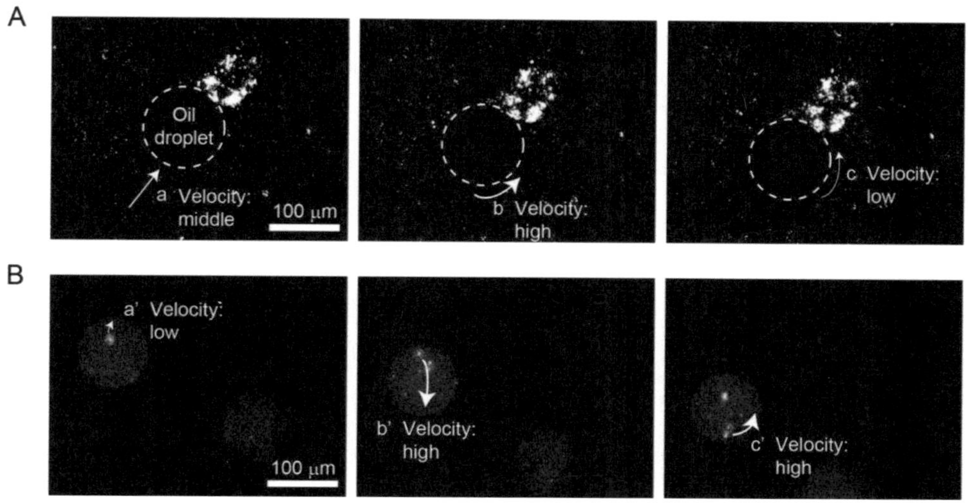

Figure 6. Visualization of the surrounding (A, time interval: 2 s) and internal (B, time interval: 3 s) flow fields of self-propelled oil droplet by using fluorescent beads under a fluorescence microscope.

2.3. Unique dynamics of micrometer-sized oil droplets in emulsion

From the proposed mechanism of self-propelled oil droplets based on the heterogeneity in the droplet surface, their motion time, direction, and mode may be controlled and altered by the molecular conversion of oil and surfactant components. Various organic reactions, such as polymerizations [14, 15], enzymatic reactions [16], and synthetic organic reactions [17–19], occur in emulsion systems because of an increase in reagent compatibility, the enhancement of reaction rates, and the induction of regioselectivity. In this section, we introduce the oil droplet system that induces unique dynamics, such as directional motion, division, and deformation, triggered by the molecular conversion of oil and surfactant components.

2.3.1. Guided directional motion

Micrometer-sized objects exhibiting well-controlled motion have drawn much attention because of their potential use as probes or sensors for exploring environmental or biological systems and as carriers for transporting compounds in very small spaces [20–22]. To control the motion time and direction of micrometer-sized oil droplets, gemini-type cationic surfactants containing a carbonate linkage in the linker moiety (**13**), the hydrolytic rate of which can be controlled by pH, have been designed and synthesized [23]. Oil droplets composed of benzaldehyde **9** exhibited self-propelled motion in an aqueous solution of pH 13 prepared by using **13** and sodium hydroxide. Their time of motion was longer than that in a basic solution of cationic surfactant without a carbonate linkage. Moreover, in a solution of **13**, oil droplets at rest underwent unidirectional self-propelled motion in a pH gradient field toward a higher concentration of sodium hydroxide. Even though they stopped within several seconds, they restarted in the same direction. Gemini-type surfactant **13** was gradually hydrolyzed, accompanied by the evolution of carbon dioxide under basic conditions to produce a pair of the corresponding monomeric surfactants **14**, which exhibit interfacial properties different from **13**. The prolonged and restart motion of the oil droplets was explained by the increase in the heterogeneity of the interfacial tension of the oil droplets. Therefore, this system where a surfactant generates other surfactants with different surface activities from the original, in an aqueous phase, could be one effective strategy for controlling the motion time and direction of oil droplets (**Figure 7**).

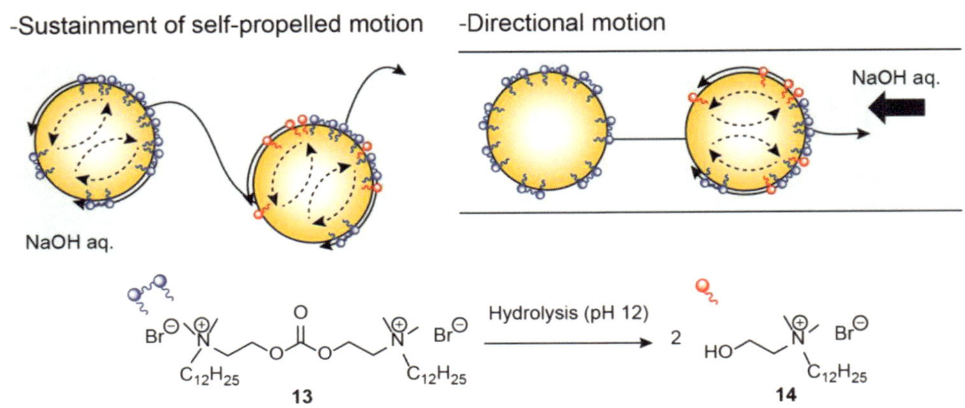

Figure 7. Schematic illustration showing the control of the motion time and direction of oil droplets in the presence of gemini-type cationic surfactant containing a carbonate linkage in the linker moiety under a basic condition.

2.3.2. Division

Even though oil droplets composed of benzaldehyde **9** exhibited self-propelled motion, they slowed down and eventually stopped because the system reached an equilibrium state in which interactions between oil and surfactant molecules were not sufficient for the induction of convective flows. Thus, the division of oil droplets during self-propelled motion is expected to occur through changes in the interaction between the components, such as the dispersion force and the ion-dipole interaction [24]. To induce such changes inside the oil droplet, 1-decanol (**15**) was added to the oil phase of **9**. Under a relatively high acidic condition, such oil droplet (mother droplet) exhibited not only self-propelled motion but also divided in the C16TAB solution. While the larger droplet (older daughter droplet) immediately self-propelled, several dozen smaller droplets (younger daughter droplets) floated in solution for approximately 1 min and then began to move. However, they were self-propelled and eventually dissolved away because of their solubilization in the bulk surfactant solution (**Figure 8A**). The self-propelled older daughter droplet also divided, and then the new older daughter droplet was immediately self-propelled. After the division, these dynamics were repeated until the older daughter droplet was dissolved away in the solution. In this emulsion, the dehydrocondensation reaction of benzaldehyde **9** and alcohol **15** occurred gradually to produce an acetal molecule, 4-(heptyloxy)-1-didecyloxybenzene (**16**). The oil droplets of **16** were not self-propelled immediately, but several of them began to move after a few minutes in the same acidic C16TAB solution. In addition, acetal **16** was completely hydrolyzed within a few minutes and was not regenerated thereafter in the dispersion. These results indicate that acetal **16** was not the primary component for self-propelled motion and that a sufficient amount of benzaldehyde **9** was required for the induction of self-propelled motion. Therefore, such division during self-propelled motion is owing to the time-course change of the oil droplet component in the reversible reaction system [25].

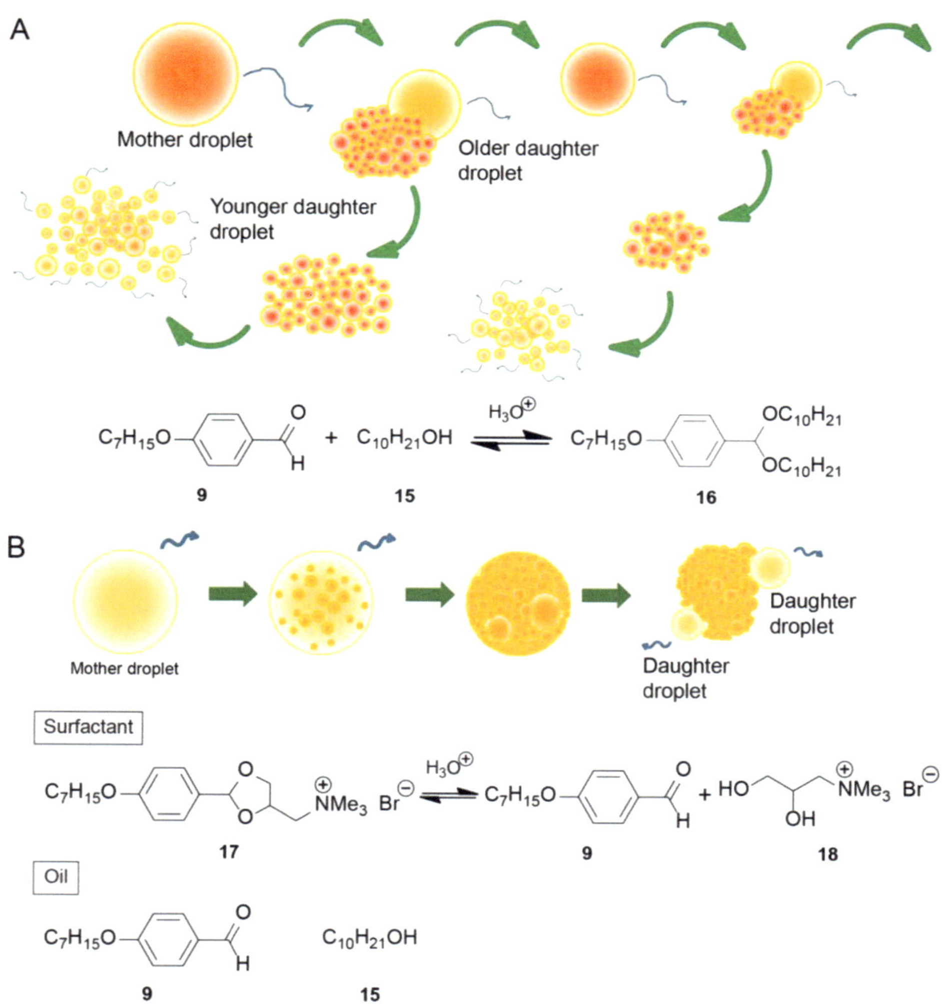

Figure 8. Schematic illustration of division during self-propelled motion of oil droplets induced by the reversible reaction of benzaldehyde **9** and alcohol **15** (A) and the hydrolysis of surfactant **17** to feed "fresh" benzaldehyde **9** into the self-propelled droplets (B).

A different division mode from the above system was also investigated using a cationic surfactant having a five-membered acetal moiety (**17**) that is hydrolyzed and affords a "fresh" benzaldehyde component (**9**) for the self-propelled oil droplets composed of benzaldehyde **9** and alcohol **15** (**Figure 8B**). In a solution composed of C16TAB and **17** containing HCl, the self-propelled motion of the initial oil droplet (mother droplet) continued until the droplet began to slow down, stopped, and then divided to release a small oil droplet (daughter droplet) which immediately self-propelled. Subsequently, other small oil droplets (daughter droplets) began to self-propel, and the daughter droplets continued in self-propelled motion. Surfactant **17** was hydrolyzed rapidly to produce benzaldehyde **9** and electrolyte **18**. On the other hand, acetal **16** was not generated under these acidic conditions. The molar fractions of **9**, **16**, and **17** did not show any change and almost reached equilibrium when the self-propelled oil droplets typically exhibited their division (ca. 10 min after the preparation of the emulsion solution), implying that the division of the self-propelled oil droplets was primarily derived from the supplied **9** and **18** (with C16TAB). Thus, the mechanism for division dynamics was proposed as follows. Self-propelled oil droplets took up the fresh benzaldehyde component (**9**) fed into them through hydrolysis of surfactant **17**, which led to the hypertrophy of oil droplets within themselves. Then, as the daughter droplets grew to reach a size that was sufficient for exhibiting the same self-propelling properties, they were released from the mother droplet [26].

2.3.3. Deformation

Although the above oil droplets maintained spherical morphology during self-propelled motion, there have been no reports on micrometer-sized self-propelled oil droplets mimicking amoeboid motion in aqueous solution. Similar to white blood cells, amoeboid motion of amoebae such as *Dictyostelium discoideum* [27] is one mode of cell locomotion associated with variations in body shape. Hence, micrometer-sized self-propelled oil droplets exhibiting such deformation could be applied as underwater carriers or probes in very narrow spaces and through many obstacles, similar to white blood cells invading blood vessels to attack tumors beneath the vessel.

To construct an oil droplet system exhibiting deformation, a mixture of two miscible compounds having similar molecular structure, fatty aldehyde **19** and fatty alcohol **15**, were used. The attraction between **19** and **15** is expected to induce spatial heterogeneity in the oil droplets. In addition, these compounds can generate acetal **20** under acidic conditions. Therefore, when these oil droplets were dispersed in a C16TAB solution, the heterogeneity resulted in differences in the adsorption and/or dissolution of the surfactant molecules at the surface and core of the oil droplets. When the molar ratio of aldehyde **19**/alcohol **15** was 60/40 mol%, self-propelled motion of the oil droplets was observed, and their shapes changed autonomously as shown in **Figure 9**. By using water-dispersed fluorescent microspheres with a diameter of 1 μm, the flow fields around self-propelled oil droplets exhibiting deformation were visualized, and their velocity was analyzed. The fluorescent microspheres at the back of the oil droplet flowed in the direction opposite to that of the oil droplet motion and approximately three times faster than those at the front of the oil droplets. The individual oil components, **19** and **15**, were liquid; however, their mixture at the molar ratio used became solid several minutes after mixing. This was likely owing to the strong attractive and non-covalent bond

http://dx.doi.org/10.5772/67249

forces between aldehyde **19** and alcohol **15**, such as hydrogen bonding and dispersion forces. On the other hand, though acetal **20** was produced in C16TAB solution containing 0.001–0.1 M HCl, its high production inhibited the induction of deformation dynamics. Therefore, there were two factors contributing to the deformation during self-propelled motion of oil droplets. These were the formation of a precipitous gradient field of interfacial tension at the back of the droplet surface and the localization of convective rolls inside the droplet caused by both the localized interfacial tension gradient and the spatiotemporal switching of intermolecular interactions among the surfactant and oil molecules [28].

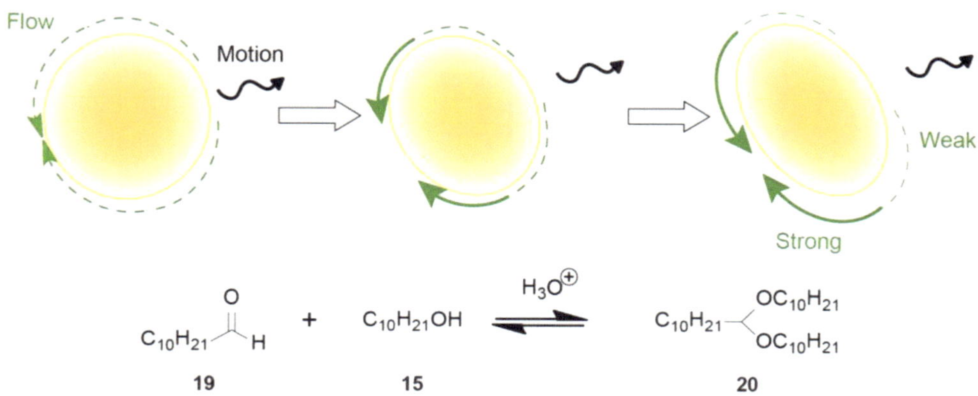

Figure 9. Schematic illustration of deformable self-propelled oil droplets composed of a fatty aldehyde and an alcohol.

3. Conclusions and future remarks

In this chapter, we demonstrate the observation methods for the self-propelled motion of micrometer-sized oil droplets using phase-contrast, polarized, and fluorescence microscopes and discuss the mechanism of motion. Since it is visible to human eye, self-propelled motion of the millimeter-sized oil droplets on the surface of aqueous solutions has been reported by many researchers [24, 29–35]. Recently, their divisions, fusions, and morphological changes have also been described. For example, Sumino et al. reported, with small-angle X-ray diffraction, that the gel phase formed spontaneously around millimeter-sized self-propelled oil droplets exhibiting "blebbing" [35]. However, since the generated self-assemblies in micrometer-sized droplet systems are difficult to identify by spectroscopic methods, the mechanisms of their self-propelled motion have not been clarified. When the mechanisms are fully understood at the nanometer to micrometer scale, the described findings will be useful to the development of more stable emulsion systems. In addition, such systems could be utilized for the reaction (microreactors) and transportation (microcarriers) of biomolecules and low-molecular-weight organic compounds [36, 37]. Furthermore, microchannel technology combined with the self-propelled droplet system is expected to lead to the development of droplet-type analysis system at the micrometer scale that can carry out reaction, analysis, and detection automatically without the need for an external force. We envisage that such automatic reaction field and analysis systems will be helpful for the improvement of novel emulsion technology.

Acknowledgements

This work was supported by the Grants-in-Aid for Scientific Research (No. 25790033 and No. 16K17504 for T.B.) and Scientific Research on Innovative Areas "Fluctuation and Structure" (No. 25103009 for T.T.) from the Ministry of Education, Culture, Sports, Science and Technology, Japan.

Author details

Taisuke Banno[1]*, Taro Toyota[2] and Kouichi Asakura[1]

*Address all correspondence to: banno@applc.keio.ac.jp

1 Department of Applied Chemistry, Faculty of Science and Technology, Keio University, Japan

2 Department of Basic Science, Graduate School of Arts and Sciences, The University of Tokyo, Japan

References

[1] Israelachivili JN. Intermolecular and Surface Forces. 3rd ed. Massachusetts: Academic Press; 2011.

[2] Abe M, Yajima I, Ogino K, Tsubaki N, Nakano Y. Properties of O/W–type microemulsion with dimyristoylphosphatidylcholine. Shikizai. 1992;**65**:212–219. DOI: 10.4011/shikizai1937.65.212

[3] Ho CC, Goetz RJ, El-Aasser MS, Vanderhoff JW, Fowkes FM. Optical microscopy of lyotropic mesophases in dilute solutions of sodium lauryl sulfate-lauryl alcohol (or cetyl alcohol)-water systems. Langmuir. 1991;**7**:56–61. DOI: 10.1021/la00049a013

[4] Peddireddy K, Kumar P, Thutupalli S, Herminghaus S, Bahr C. Solubilization of thermotropic liquid crystal compounds in aqueous surfactant solutions. Langmuir. 2016;**28**:12426–12431. DOI: 10.1021/la3015817

[5] Takahashi Y, Fukuyasu K, Horiuchi T, Kondo Y, Stroeve P. Photoinduced demulsification of emulsions using a photoresponsive gemini surfactant. Langmuir. 2014;**30**:41–47. DOI: 10.1021/la4034782

[6] Toyota T, Tsuha H, Yamada K, Takakura K, Ikegami T, Sugawara T. *Listeria*-like motion of oil droplets. Chemistry Letters. 2006;**35**:708–709. DOI: 10.1246/cl.2006.708

[7] Hanczyc MM, Toyota T, Ikegami T, Packard N, Sugawara T. Fatty acid chemistry at the oil–water interface: Self-propelled oil droplets. Journal of the American Chemical Society. 2007;**129**:9386–9391. DOI: 10.1021/ja0706955

[8] Toyota T, Maru N, Hanczyc MM, Ikegami T, Sugawara T. Self-propelled oil droplets consuming "fuel" surfactant. Journal of the American Chemical Society. 2009;**131**:5012–5013. DOI: 10.1021/ja806689p

[9] Banno T, Kuroha R, Toyota T. pH-Sensitive self-propelled motion of oil droplets in the presence of cationic surfactants containing hydrolyzable ester linkages. Langmuir. 2012;**28**:1190–1195. DOI: 10.1021/la2045338

[10] Banno T, Miura S, Kuroha R, Toyota T. Mode changes associated with oil droplet movement in solutions of gemini cationic surfactants. Langmuir. 2013;**29**:7689–7696. DOI: 10.1021/la401416h

[11] Reiss-Husson F, Luzzati V. The structure of the micellar solutions of some amphiphilic compounds in pure water as determined by absolute small-angle X-ray scattering techniques. The Journal of Physical Chemistry. 1964;**68**:3504–3511. DOI: 10.1021/j100794a011

[12] Wilcoxon JP, Kaler EW. Statics and dynamics of concentration fluctuations in nonionic micellar solutions. Journal of Chemical Physics. 1987;**86**:4684–4691. DOI: 10.1063/1.452709

[13] Maass CC, Kruger C, Herminghaus S, Bahr C. Swimming droplets. Annual Review of Condensed Matter Physics. 2016; **7**:171–193. DOI: 10.1146/annurev-conmatphys-031115-011517

[14] Lade O, Beizai K, Sottmann T, Strey R. Polymerizable nonionic microemulsions: Phase behavior of H_2O-*n*-alkyl methacrylate-*n*-alkyl poly(ethylene glycol) ether (CiEj). Langmuir. 2000;**16**:4122–4130. DOI: 10.1021/la991232i

[15] Gao F, Ho CC, Co CC. Sugar-based microemulsion glass templates. Journal of the American Chemical Society. 2004;**126**:12746–12747. DOI: 10.1021/ja0467115

[16] Orlich B, Schomäcker R. *Candida rugosa* lipase reactions in nonionic w/o-microemulsion with a technical surfactant. Enzyme and Microbial Technology. 2001;**28**:42–48. DOI: 10.1016/S0141-0229(00)00268-4

[17] Holmberg K. Organic reactions in microemulsions. Current Opinion in Colloid and Interface Science. 2003;**8**:187–196. DOI: 10.1016/S1359-0294(03)00017-7

[18] López-Quintela MA, Tojo C, Blanco MC, García LR, Leis JR. Microemulsion dynamics and reactions in microemulsions. Current Opinion in Colloid and Interface Science. 2004;**9**:264–278. DOI: 10.1016/j.cocis.2004.05.029

[19] Kobayashi S, Manabe K. Development of novel Lewis acid catalysts for selective organic reactions in aqueous media. Accounts of Chemical Research. 2002;**35**:209–217. DOI: 10.1021/ar000145a

[20] Shoji M, Ikuta K, Korogi H. Submicron manipulation tools driven by light in a liquid. Applied Physics Letters. 2003;**82**: 133–135. DOI: 10.1063/1.1533853

[21] Uchida E, Azumi R, Norikane Y. Light-induced crawling of crystals on a glass surface. Nature Communications. 2015;**6**:7310. DOI: 10.1038/ncomms8310

[22] Paxton WF, Sundararajan S, Mallouk TE, Sen A. Chemical locomotion. Angewandte Chemie International Edition. 2006;**45**: 5420–5429. DOI: 10.1002/anie.200600060

[23] Miura S, Banno T, Tonooka T, Osaki T, Takeuchi S, Toyota T. pH-Induced motion control of self-propelled oil droplets using a hydrolyzable gemini cationic surfactant. Langmuir. 2014;**30**:7977–7985. DOI: 10.1021/la5018032

[24] Browne KP, Walker DA, Bishop KJM, Grzybowski BA. Self-division of macroscopic droplets: Partitioning of nanosized cargo into nanoscale micelles. Angewandte Chemie International Edition. 2010;**49**:6756–6759. DOI: 10.1002/ange.201002551

[25] Banno T, Kuroha R, Miura S, Toyota T. Multiple-division of self-propelled oil droplets through acetal formation. Soft Matter. 2015;**11**:1459–1463. DOI: 10.1039/C4SM02631D

[26] Banno T, Toyota T. Molecular system for the division of self-propelled oil droplets by component feeding. Langmuir. 2015;**31**:6943–6947. DOI: 10.1021/acs.langmuir.5b00904

[27] Bray D. Cell Movements. 2nd ed. New York: Garland Publishing; 2001.

[28] Banno T, Asami A, Ueno N, Kitahata H, Koyano Y, Asakura K, Toyota T. Deformable self-propelled micro-objects comprising underwater oil droplets. Scientific Reports. 2016;**6**:31292. DOI: 10.1038/srep31292

[29] Florea L, Wagner K, Wagner P, Wallace GG, Benito-Lopez F, Officer DL, Diamond D. Photo-chemopropulsion—light-stimulated movement of microdroplets. Advanced Materials. 2014:**26**:7339–7345. DOI: 10.1002/adma.201403007

[30] Čejková J, Novák M, Štěpánek F, Hanczyc MM. Dynamics of chemotactic droplets in salt concentration gradients. Langmuir. 2014:**30**:11937–11944. DOI: 10.1021/la502624f

[31] Pimienta V, Brost M, Kovalchuk N, Bresch N, Steinbock O. Complex shapes and dynamics of dissolving drops of dichloromethane. Angewandte Chemie International Edition. 2011;**50**:10728–10731. DOI: 10.1002/anie.201104261

[32] Nagai K, Sumino Y, Kitahata H, Yoshikawa K. Mode selection in the spontaneous motion of an alcohol droplet. Physical Review E. 2005;**71**:065301. DOI: 10.1103/PhysRevE.71.065301

[33] Lagzi I, Soh S, Wesson PJ, Browne KP, Grzybowski BA. Maze solving by chemotactic droplets. Journal of the American Chemical Society. 2010;**132**:1198–1199. DOI: 10.1021/ja9076793

[34] Caschera F, Rasmussen S, Hanczyc MM. An oil droplet division–fusion cycle. ChemPlusChem. 2013;78:52–54. DOI: 10.1002/cplu.201200275

[35] Sumino Y, Kitahata H, Seto H, Nakata S, Yoshikawa K. Spontaneous deformation of an oil droplet induced by the cooperative transport of cationic and anionic surfactants through the interface. The Journal of Physical Chemistry B. 2009;**113**:15709–15714. DOI: 10.1021/jp9037733

[36] Cygan ZT, Cabral JT, Beers KL, Amis EJ. Microfluidic platform for the generation of organic-phase microreactors. Langmuir. 2005;**21**:3629–3634. DOI: 10.1021/la0471137

[37] Ji J, Zhao Y, Guo L, Liu B, Jib C, Yang P. Interfacial organic synthesis in a simple droplet-based microfluidic system. Lab on a Chip. 2012;**12**:1373–1377. DOI: 10.1039/C2LC40052A

Chapter 8

Influence of Amphiphiles on Percolation of AOT-Based Microemulsions Prediction Using Artificial Neural Networks

Gonzalo Astray, Antonio Cid,
Oscar Adrián Moldes and Juan Carlos Mejuto

Additional information is available at the end of the chapter

http://dx.doi.org/10.5772/66766

Abstract

In this chapter, the ability of artificial neural networks was evaluated to predict the influence of amphiphiles as additive upon the electrical percolation of dioctyl sodium sulfosuccinate (AOT)/isooctane/water microemulsions. In particular, water/AOT/isooctane microemulsion behaviour has been modelled. These microemulsions have been developed in presence of 1-n-alcohols, 2-n-alcohols, n-alkylamines and n-alkyl acids. In all cases, a neural network has been obtained to predict with accuracy the experimental behaviour to identify the physico-chemical variables (such as additive concentration, molecular mass, log P, pK_a or chain length) that exert a greater influence on the model. All models are valuable tools to evaluate the percolation temperature for AOT-based microemulsions.

Keywords: percolation, microemulsion, AOT, additives, modelling, artificial neural network

1. Introduction

Microemulsions are colloidal self-organized systems, composed of a polar phase, in our case water, and a non-polar phase, isooctane, stabilized by a surfactant film that causes the formation of droplets of the dispersed phase in the continuous phase. In our case, the surfactant used was the AOT (dioctyl sodium sulfosuccinate) whose main advantage is the formation of a stable microemulsion in wide concentration ranges. Actually, this kind of microemulsion is known as water in oil (w/o), that is, water is the dispersed phase and continuous phase will be the apolar medium.

Internal dynamics of microemulsions has been largely studied, especially on the phenomenon of electrical percolation [1–4]. Electrical percolation is characterized by an increment in electrical conductivity when the temperature, or the volume fraction of the dispersed phase, reaches a critical value [5]. In this sense, the change in electrical conductivity is very characteristic, with variations from small values to large values, which is the typical behaviour of small droplets dispersed in a non-conductive continuous medium [5, 6].

Relationship between electrical percolation and constant rates was demonstrated by Lang and co-workers [7–9], and they showed that the exchange of materials between droplets has influence on the rate of fast chemical reactions in w/o microemulsions [5]. Mathew et al. [10] observed that percolation threshold is altered by small additives concentrations such as cholesterol or gramicidin [5]. These findings have been confirmed by literature during the last decade [11–14]. In support of this, we can say that percolation is not a consequence of bicontinuous structures presented in the medium, because the structure of discrete droplets is not changed [5]. When percolation threshold is approached, the number of collisions presents a huge increment, leading to the formation of droplet clusters with interdroplet channels that allow transport of ions, giving rise to an increase in conductivity [5].

In the last decade, our research group has studied the effects of different additives on the electrical conductivity, and other properties, for water/AOT/isooctane microemulsions (aerosol OT or dioctyl sodium sulfosuccinate, isooctane and water) [5, 15–22]. The influence of different additives was explained on the basis of changes in the surfactant film structure and different solubility of the complex system. The manuscript shows the artificial neural networks (ANNs) as a valuable tool to predict percolation threshold for microemulsions (AOT/isooctane/water) in the presence of different amphiphiles, because there are no mathematical tools to predict the influence of additives on the internal dynamics of microemulsions. The different additives were molecules with amphiphilic character composed of a variable apolar hydrocarbon chain, with a polar head group. In particular, the effect of 1-n-alcohols, 2-n-alcohols, n-alkylamines and n-alkyl acids was modelled. The effects of these compounds have been previously described in the literature (*vide infra*).

In the last two or three decades, artificial neural networks have become one of the most applied methodologies to develop models for non-linear behaviours [23–25]. ANNs are a mathematical method that tries to imitate the reasoning of human brain [26]. Individual units, called neurons, form neural models that are the fundamental unit to model complex problems [24]. For this reason, neural models are being applied in different areas of study, such as (i) hydrology to predict the discharge of rivers and prevent floods and water loggings in spite of the large number of variables involved in the process [25, 27], (ii) chemistry to model the infinite dilution activity coefficients of halogenated hydrocarbons that provide important information about the solute-solvent interactions [28], (iii) energy science to model wind speed which is important for renewable energy and energy market efficiency [29], (iv) biorefinery to determine ideal conditions to obtain new oligosaccharide mixtures production from sugar beet pulp [24], or, even, (v) business, management and accounting to predict overall bank customer satisfaction and to prioritize factors for customer satisfaction [30], inter alia.

In our research group, we use a multi-layer perceptron (MLP), which can model complex non-linear relationship between independent and dependent variables. This kind of ANNs is one of the most used neural models in the literature [23, 31, 32].

2. Materials and methods

2.1. Percolation temperature determination

An experimental procedure to determine percolation temperature has been described previously [20–22]. A Crison GPL 32 conductivity meter was used. Percolation threshold (T_p) can be represented using the method described by Kim et al. [33], or using the sigmoidal Boltzmann equation (SBE), suggested by Moulik et al. [34].

$$log\kappa = log\,\kappa_i\left[1+\left(\frac{log\,\kappa_i - log\,\kappa_f}{log\,\kappa_f}\right)\left\{1+\frac{e^{(T-T_p)}}{\Delta T}\right\}^{-1}\right] \quad (1)$$

In Eq. (**1**), k is the conductance, T represents temperature and ΔT is the constant range of temperatures. In the equation, the different subscripts i, f and p represent the initial state, final state and percolation threshold, respectively. Percolation temperatures obtained by both methods [33–34] are compatible, as we can see in **Figure 1**. T_p obtained from the SBE [34] will be used in the discussion.

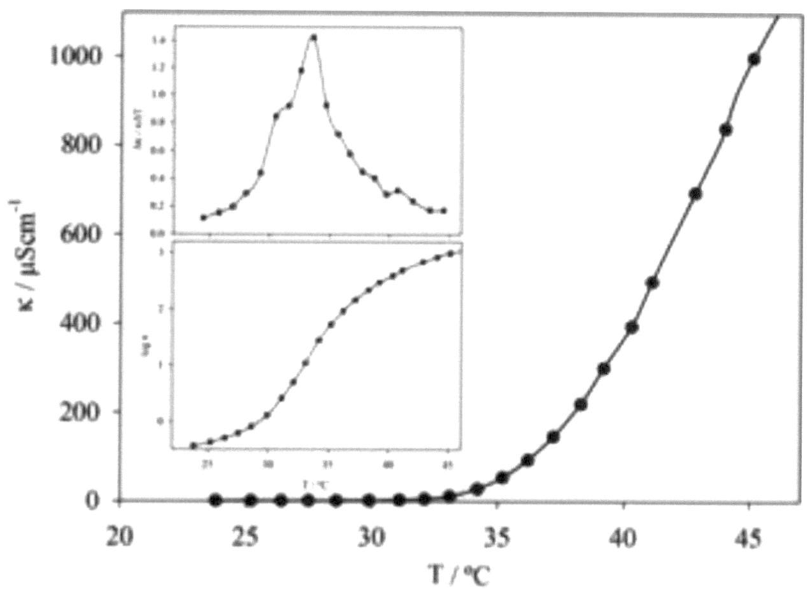

Figure 1. Determination of percolation threshold for water in oil AOT-based microemulsion (AOT/iC_8/H_2O) for both Kim et al. method and Moulik et al. method [34]. Lower insert corresponds with SBE method and upper insert stands the graphical approach. Microemulsion composition: [AOT] = 0.5 M, W = [H_2O]/[AOT] = 22.

2.2. Artificial neural networks

The ANN procedure starts choosing different groups of data. The first group, called training group, is formed by the training cases used to develop the neural model. The second group, called validation group, is formed by reserved cases used to validate the model. When two groups are selected, neural models must be developed using trial and error technique to determine the best configuration parameters (weights and bias values) and the best model topology to predict the desired variable [24, 27, 35, 36].

The training cases are presented to the first layer, called input layer, which is formed by different neurons to receive the input information. This information is presented as an input vector (Eq. (**2**)), and it is propagated using a specific function, called propagation function, from input layer to the first intermediate layer where learning process occurs (Eq. (**3**)), and then to the final layer [24, 27, 35, 36]. This equation is implemented in each intermediate and output neurons. Propagation function converts all input information into one signal response (S). The input values (x_i) are processed with the neuron weight (w_{ik}) linking the intermediate neuron (k) with the previous neuron (i), and added to the bias value associated to neuron k (**Figure 2**).

$$x = \left(x_1, x_2, x_3, \ldots, x_n\right) \tag{2}$$

$$S = \sum_{i=1}^{M} w_{ik} x_i + b_k \tag{3}$$

The single data (S) is treated by an activation function that provides an output neuron signal (y_k). This procedure is performed in each neural neuron to obtain a final value (predicted value ($y_{predicted}$)). There are different activation functions, but the most used activation function is the logistic function [24, 36–39], which is the function used in all the studies conducted by our research group (Eq. (**4**)).

$$y_k = \frac{1}{1 + e^S} \tag{4}$$

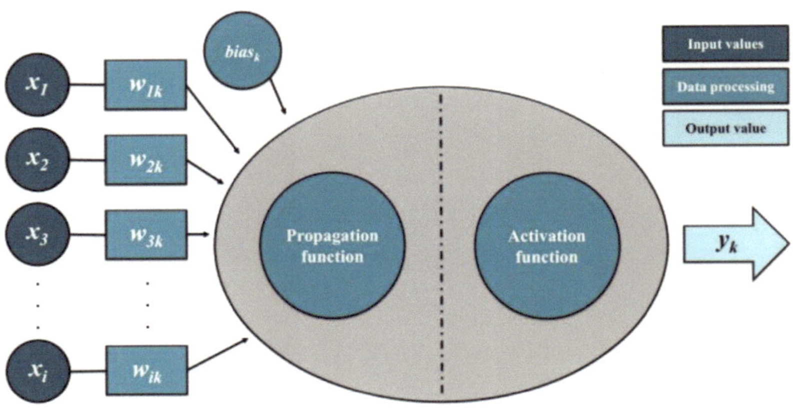

Figure 2. Artificial neuron operating procedure.

To facilitate the model topology identification, in our research group we use the following terminology [24, 27, 35]: N_{in}, N_{k-1}, N_{k-2}, N_{k-3} and N_{out}, where N_{in} and N_{out} represent the neurons in the input and output layers, respectively. N_{k-1}, N_{k-2} and N_{k-3} are the neurons in the first, second and third hidden layers, respectively. In **Figure 3**, we can see an example of neural models with five neurons in the input layer, three intermediate neurons in the next layer and only one output neuron.

2.3. Neural power prediction

Neural models learn from the training cases and generalize the acquired knowledge to validation cases, which provide an idea of neural model power prediction. This power prediction must be checked using different adjustment parameters [24, 27, 35]; in our research group, we usually used (i) the determination coefficient (R^2) (Eq. (**5**)) and (ii) the root-mean-squared error (RMSE) (Eq. (**6**)). As we know, good determination coefficient is necessary, a small error between real data and predicted data is also necessary [24]; for this reason, RMSE must be checked to know how close to zero is the model error. The best models have been chosen based on the lowest RMSE in the validation phase.

$$R^2 = \frac{\mathrm{SSR}}{\mathrm{SST}} = \frac{\sum_{i=1}^{N}\left(y_{\mathrm{predicted}} - \bar{y}\right)^2}{\sum_{i=1}^{N}\left(y_{\mathrm{real}} - \bar{y}\right)^2} \tag{5}$$

$$\mathrm{RMSE} = \sqrt{\frac{\sum_{i=1}^{N}\left(y_{\mathrm{predicted}} - y_{\mathrm{real}}\right)^2}{N}} \tag{6}$$

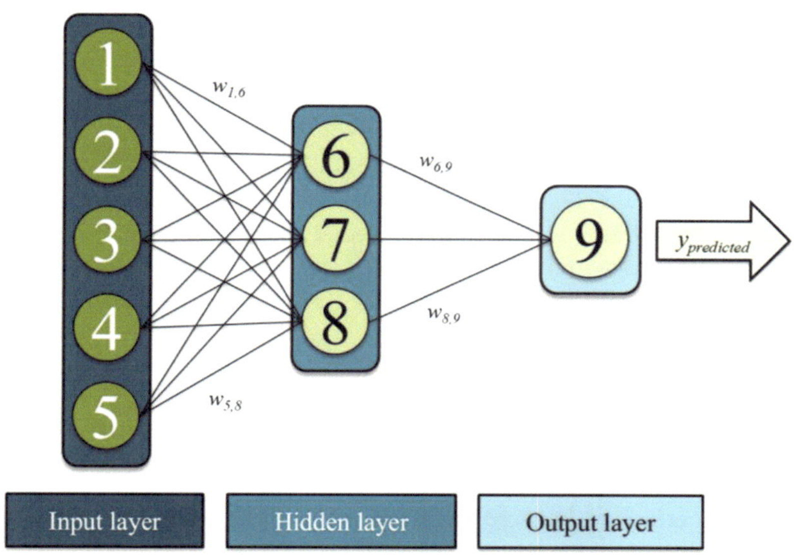

Figure 3. Example of neural model topology 5-3-1 with five neurons in the input layer, three neurons in intermediate layers and one neuron in the output layer.

2.4. Equipment and software

All models have been implemented in two servers specifically designed with a client-server architecture. Clients running virtual machines optimized for peak performance implementation. ANN models were developed using EasyNN plus (from Neural Planner Software Ltd.), and data were fitted, and plotted, using commercial software (Microsoft Excel from Microsoft, USA) and Sigmaplot Trial versions, respectively. The figures were developed using Power Point from Microsoft. Geometrical parameters of amphiphiles were determined by MM2 with CS Chem Bats 3D Pro 4.0 by Cambridge Soft Corporation, based on QCPE 395 [40, 41].

3. Results

3.1. Percolation prediction in AOT-based microemulsions

Our first tests for the prediction from the influence of additives on the percolation phenomenon were performed to analyse the influence of salts on the percolation temperature of AOT-based microemulsions [42]. For this neural model, 58 cases were used [5, 43–45] in which the kind of salt, concentration and microemulsion composition were varied [42]. In these neural models, (i) *W* value of the microemulsion (W = [H2O]/[AOT]), (ii) additive concentration ([Add]), (iii) molecular weight of the additive (M_w), (iv) atomic radii of salt components (r_{atomic}) and (v) ionic radii (r_{ionic}) of salt were used as input variables [42]. In this case, the best ANN presents a topology with five input nodes, two middle layers with 11 and seven nodes and one output neuron (see **Figure 4**) [42].

The obtained root-mean-squared error was 0.18°C (R = 0.9994) for the training phase and 0.64°C (R = 0.9789) for the prediction set [42]. It should be emphasized that power predictions for all salt families were satisfactory (see **Figure 5**) [42]. Comparing the R values for lithium, sodium and potassium salts, we can see that these values were 0.8997, 0.9993 and 0.9970, respectively [42]. Anion analysis shows that the correlation coefficients were 0.9953, 0.9971, 0.9989, 0.9762 and 0.9712 for the different salts, fluorides, chlorides, bromides, iodides and perchlorates, respectively [42]. In the developed model, there are two cases having a significant deviation (cases with HCl as additive) [42]. Experimental and predicted values for these two cases present standard deviations of 1.01 and 1.09°C, which correspond with errors below 3% [42].

Since the efficiency of ANNs to predict the influence of salts on the electrical percolation has been demonstrated, we have addressed the possibility of extending our studies to other additives.

Our laboratory has conducted extensive studies on the effect of additives on the internal dynamics of microemulsions in recent years [46]; so, the next step was the application of ANNs on the systems in which the additives were small organic molecules, particularly ureas and thioureas [47]. In this research, the developed ANN model presents a topology with three input nodes, one hidden layer composed for two neurons and one node in the output layer [47]. ANN model presents a correlation coefficient of 0.9251 for the

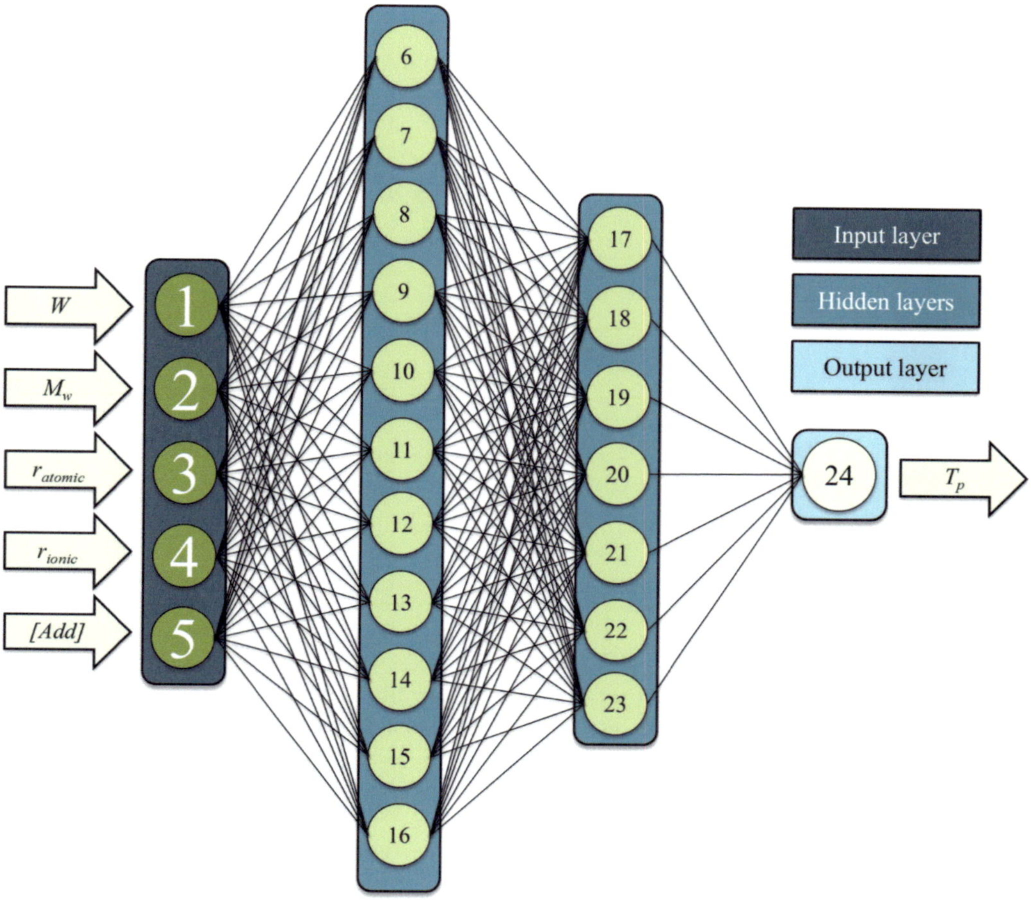

Figure 4. ANN architecture for salt influence prediction upon AOT-based microemulsions percolation. Modified with permission from Cid et al. [42].

training phase and 0.9719 for the validation phase [47] (see **Figure 6**). To develop the best model, different input variables were assayed, including (i) critical molecules volume and (ii) molecular weight, (iii) water solubility, (iv) log *P* and (v) concentration for additive component [47]. Nevertheless, some of these variables had no statistical importance to improve the model results, and therefore only (i) additive concentration, (ii) log *P* and (iii) *W* value of the microemulsion were used as input variables [47]. With these new input parameters, we can observe that the new neural model is simpler than the previous model developed for salts as additives [42]. In this case, 95 microemulsion compositions were used in the training phase and 15 microemulsion compositions were used in the validation phase [5, 47, 48]. The neural model for salts as additives present an average error equal to <0.3 and <1.4% for the training and validation phases, respectively [42], while for the model for ureas and thioureas, the error was ≈1% for the training phase and 0.9% for the validation phase [47]. In this neural model, the ANN presents bigger errors for the training and validation phases against the previous model (*vide supra*) [47].

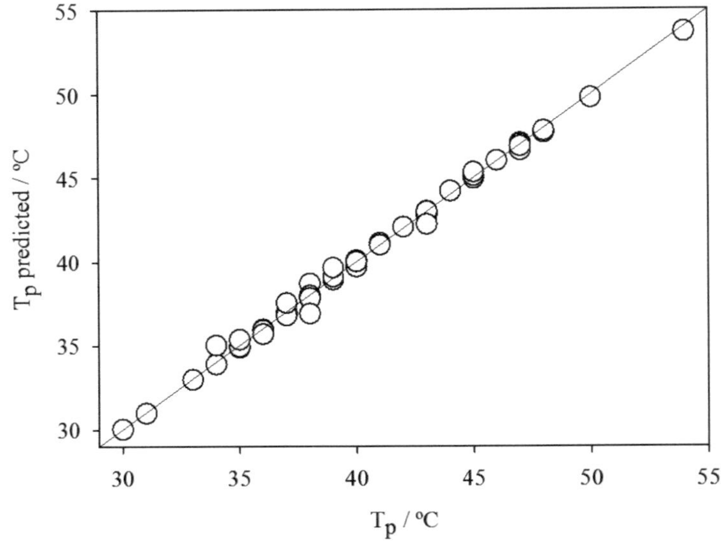

Figure 5. Experimental versus calculated value of T_p for AOT-based microemulsions with salts as additive. Modified with permission from Cid et al. [42].

Nevertheless, these results prove that our ANNs are valid predictive tools for percolative phenomena of microemulsions. In fact, satisfactory results were found for crown ethers [49–53]–both crown ethers and aza-crown ethers–glymes and polyethylene glycols [54, 55].

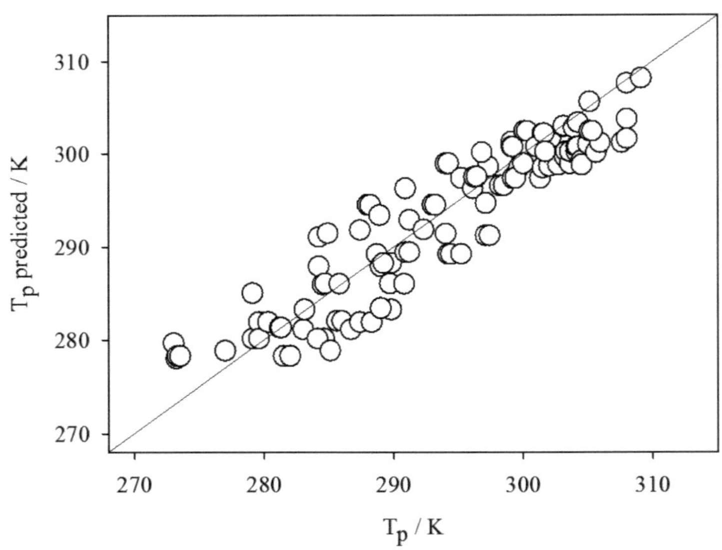

Figure 6. Experimental versus calculated value of T_p for AOT-based microemulsions with ureas and thioureas as additive. Modified with permission from Montoya et al. [47].

Previously, huge numbers of neural models had been developed to obtain a good prediction model. In this sense, the best neural model, with topology 10-8-1, presents a good root-mean-squared error around 1.169°C [49]. This error is in concordance with other neural models developed for different additives described above. Different input variables were chosen due its relationship with the nature and structure of the molecule, which influence the packing capabilities of surfactant film [49]. In this sense, the following variables: (i) additive concentration, (ii) number of atoms that conform a ring in a crown ether, (iii) number of heteroatoms, (iv) number of oxygen atoms, (v) number of nitrogen atoms, (vi) number of benzene rings in the molecule, (vii) molecular mass, (viii) log *P*, (ix) maximum number of bonds between rings and (x) minimum number of bonds between rings were used as input variables for neural model [49]. Each variable provide different information for neural model such as: (i) additive concentration provides information about the impact of different additive amounts, (ii) number of atoms, molecular mass, bond and rings provide information about structure that can affect interactions with surfactant and (iii) log *P* provides information about polarity [49]. With the use of these variables, our research group has been able to simulate the effect of four different families of structurally related compounds (crown ethers, azo-crown ethers, benzo-crown ether and dibenzo-crown ethers) [49].

For the former, two series of models were developed, one for glymes and the other for polyethylene glycols. Available datasets for glymes [55] consisted of 44 microemulsion compositions and for polyethylene glycols [55–57] consisted of 82 microemulsion compositions.

The best developed neural model to predict glymes percolation temperature presents a topology 5-5-1, that is, five nodes in input layer, five nodes in the only intermediate layer and one neuron in the output layer [55]. This neural model has been trained with 32 experimental cases, and 11 experimental cases were used to validate the neural model [55]. **Figure 7** shows a scheme of this neural model [55]. Best polyethylene glycols model presents a topology with five input neurons, three intermediate layers with eight, eight and five neurons and an output layer with one node (see **Figure 8**) [55]. This model was developed using 68 training cases and were validated with 14 experimental cases [55]. These two neural models present RMSE values of 0.19 and 0.06°C for glymes and polyethylene glycols training phases, respectively, with correlation coefficients of 0.9996 and 0.9999 [55], on the other hand, for the validation phase, the models presents RMSE values of 0.75, and 0.10°C, respectively, with correlation coefficients of 0.9938 and 0.9952 [55].

Crown ethers, glymes and polyethylene glycols are similar molecules; however, the first is characterized by being cyclic molecules and the others are linear. Beside this, crown ethers have a complex behaviour when they are used as additives in AOT microemulsions; this behaviour contrasts with glymes and polyethylene glycols. In crown ether microemulsions, percolation temperature increases slowly with concentration, nevertheless the value begins to decrease from a certain value, so higher concentrations reduce the percolation threshold. This behaviour is a combination of the two effects: (i) the microdroplet structure reinforcement by ion capture and a subsequent transference to surfactant film and (ii) the destabilization effect due to the non-polar region of the additive. On the other hand, glymes and polyethylene glycols microemulsions

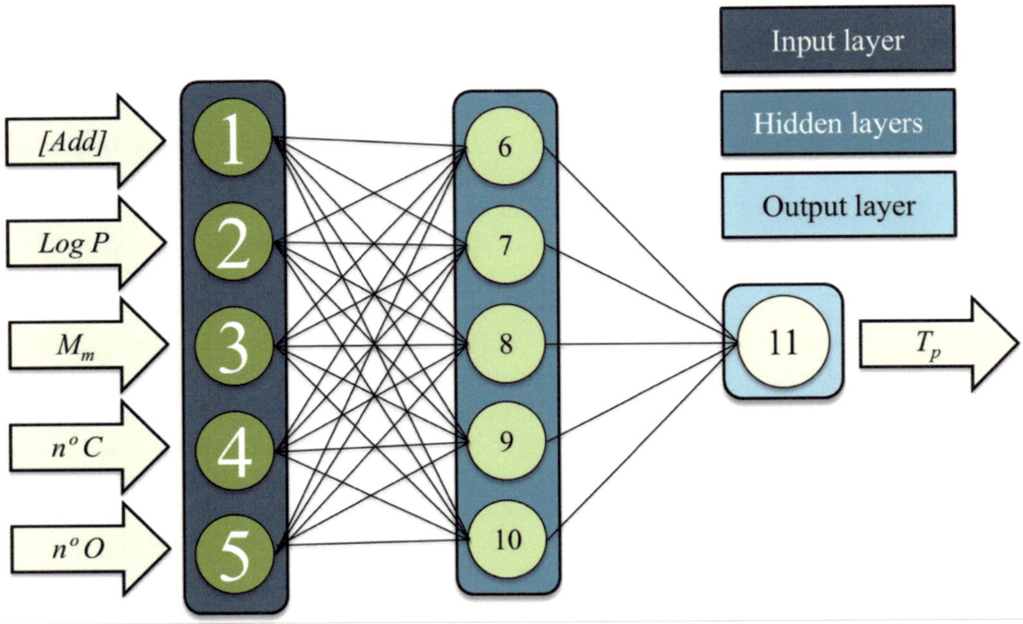

Figure 7. ANN architecture for glyme influence prediction upon AOT-based microemulsions percolation. Modified with permission from Moldes et al. [55].

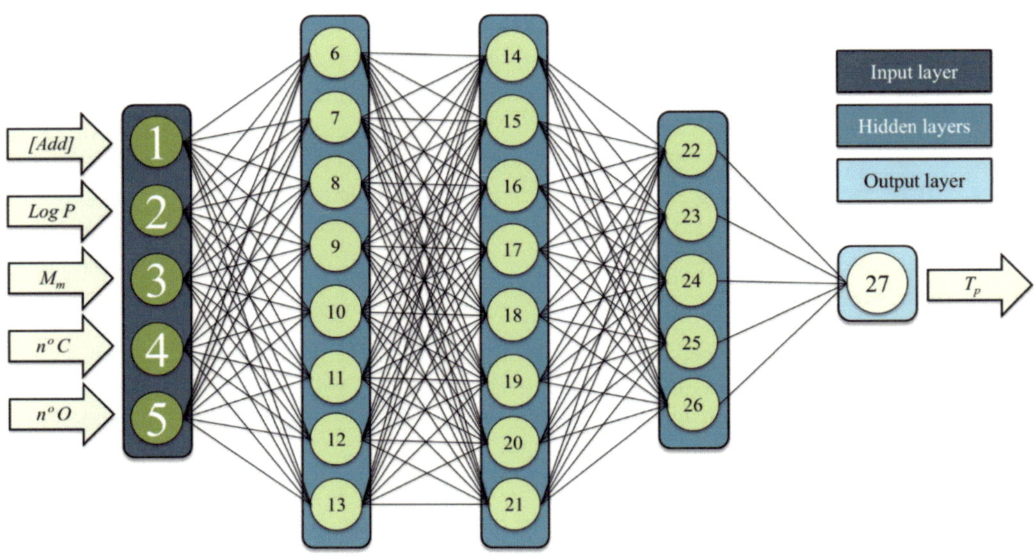

Figure 8. ANN architecture for polyethylene glycol influence prediction upon AOT-based microemulsions percolation. Modified with permission from Moldes et al. [55].

http://dx.doi.org/10.5772/66766

behaviour is simpler than crown ethers behaviour, mainly like the effect exerted by urea and other small organic molecules (*vide supra*). Due to the impossibility to develop an accurate prediction model for crown ethers, glymes and polyethylene glycols, it is necessary to limit the inputs variables to molecular descriptor and concentration in order to develop a simple model.

Another neural model has been developed for similar additives, propylene glycols, which produce a decrease in percolation temperature [58]. However, neural model cannot predict successfully the predicting percolation threshold in presence of these kinds of additives. In this sense, a single neural model for three additives is not possible yet. Even though, acceptable root-mean-squared errors were obtained with the two models described in this work.

Neural model topologies for glymes and glycols are different, as we can see above (see **Figure 7** and **Figure 8**). Neural models for glymes present just one intermediate layer, while neural model for polyethylene glycols present three intermediate layers [55]. Log *P*, which provides information of polarity for neural generalization, was not the most important input variable [55] and has been relegated to a second position by additive concentration in both models [55]. The importance of a non-dimensional parameter is obtained from the sum of importance of each input neuron with intermediate neurons [55]. Despite the difference in the order of these two variables, the importance of every input node was in a similar range for both models, which indicates similarity [55].

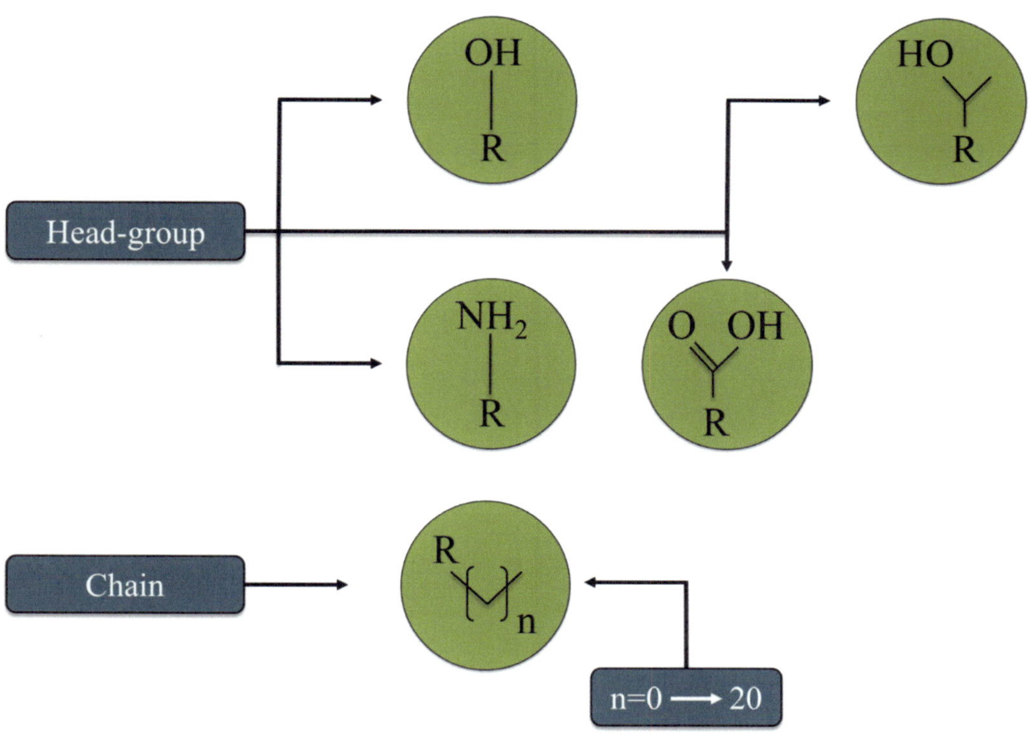

Figure 9. Amphiphiles used as additives in AOT-based microemulsions.

3.2. Percolation prediction in AOT-based microemulsions in the presence of amphiphiles

In order to evaluate the effect of amphiphiles on percolation threshold, the influence of 1-n-alcohols [59], 2-n-alcohols [59], n-alkylamines [60-63] and n-alkyl acids [64, 65] had been analysed. This allowed us to estimate the influence of the chain length of the molecule with constant head group and also the influence of the head group while the chain length remains invariant (see **Figure 9**).

The alcohols (both 1-n-alcohols and 2-n-alcohols) were modelled using the same ANN [59]. The best neural model presents a topology with five input nodes, a single intermediate layer with 11 neurons and one node in the output layer to predict percolation temperature (5-11-1) [59] (**Figure 10**). This model was developed with five input neurons: (i) additive concentration [Add], (ii) molecular weight (M_w), (iii) number of carbon atoms (n°C), (iv) pK_a, and (v) log P [59].

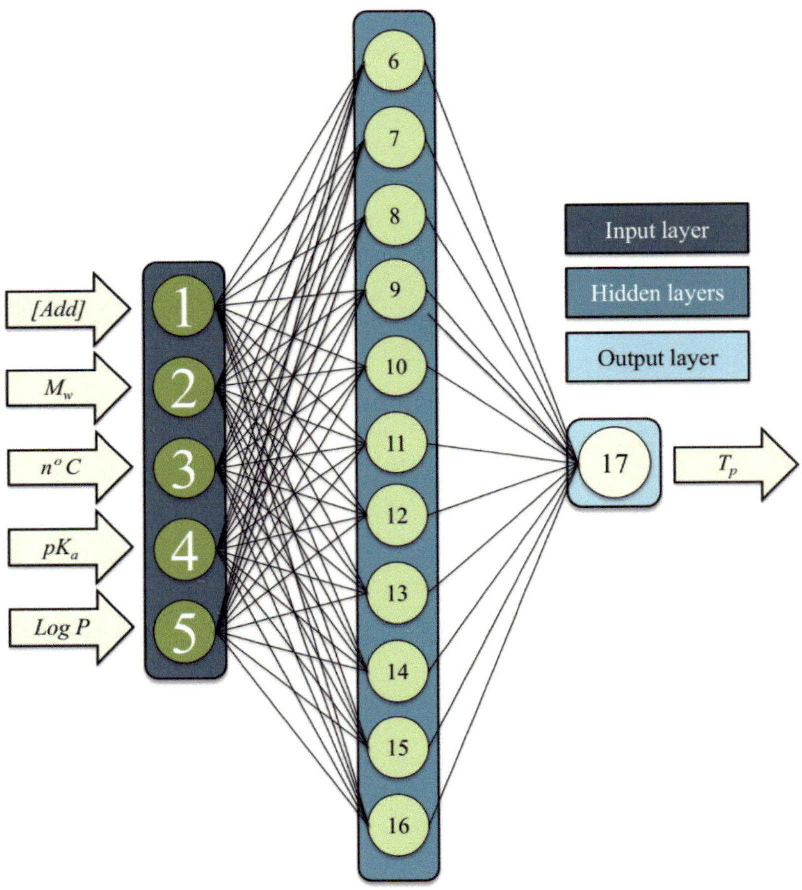

Figure 10. ANN architecture for 1-n-alcohols and 2-n-alcohols influence prediction upon AOT-based microemulsions percolation. Modified with permission from Moldes et al. [59].

This neural model was trained with 41 microemulsion compositions (67.2% of the total cases) and 20 compositions for valuation phase (32.8% of the total cases) [59]. Best neural model, 5-11-1, presents a root-mean-squared error of 0.73°C (R = 0.9939) in the training phase and 0.98°C (R =0.9869) for the validation phase [59] (see **Figure 11** and **Figure 12**).

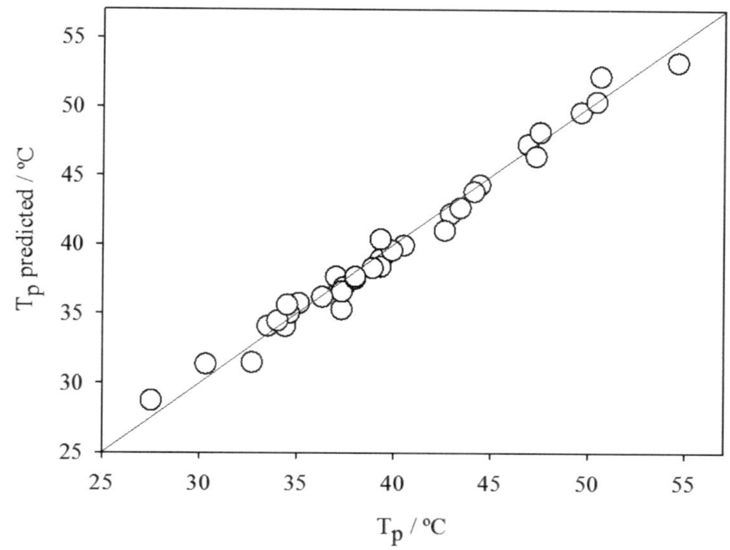

Figure 11. Experimental versus calculated value of T_p for AOT-based microemulsions with 1-n-alcohols as additive. Modified with permission from Moldes et al. [59].

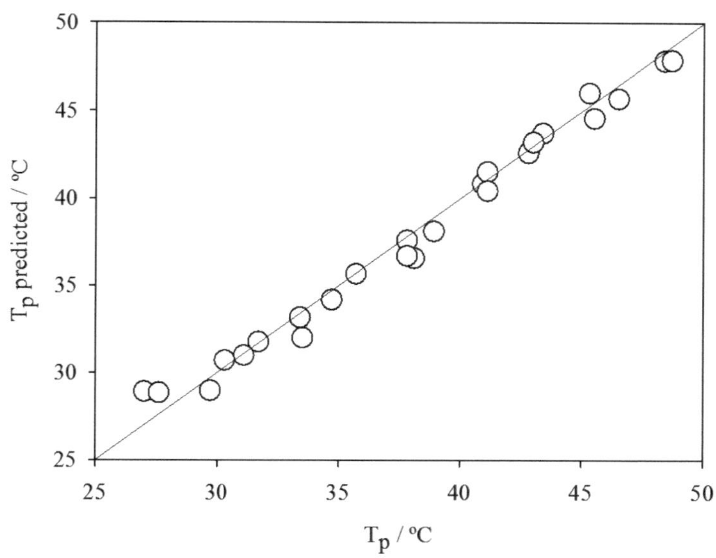

Figure 12. Experimental versus calculated value of T_p for AOT-based microemulsions with 2-n-alcohols as additive. Modified with permission from Moldes et al. [59].

In the case of carboxylic acids [65], the best neural model presents a topology with five neurons in the input layer, two intermediate layers with five and 10 neurons and an output layer with one node [65]. Related to this model, we can check that the most important variable, according to importance value, is the acid concentration, followed by log *P*, the number of carbons, chain length (these two input variables with a similar importance) and finally, pK_a[65]. Neural models present a root-mean-squared error for the training phase of 0.41°C, corresponding to a 0.72% average percentage deviation (APD) [65]. These models were checked with two different validation data groups: the first one, with similar cases used to train the model, presents an RMSE of 0.61°C (APD of 1.20%) and the second group was composed of two unknown *n*-alkyl acids (not previously trained), and presents an RMSE of 0.75°C (APD of 1.43%), (**Figure 13**) [65].

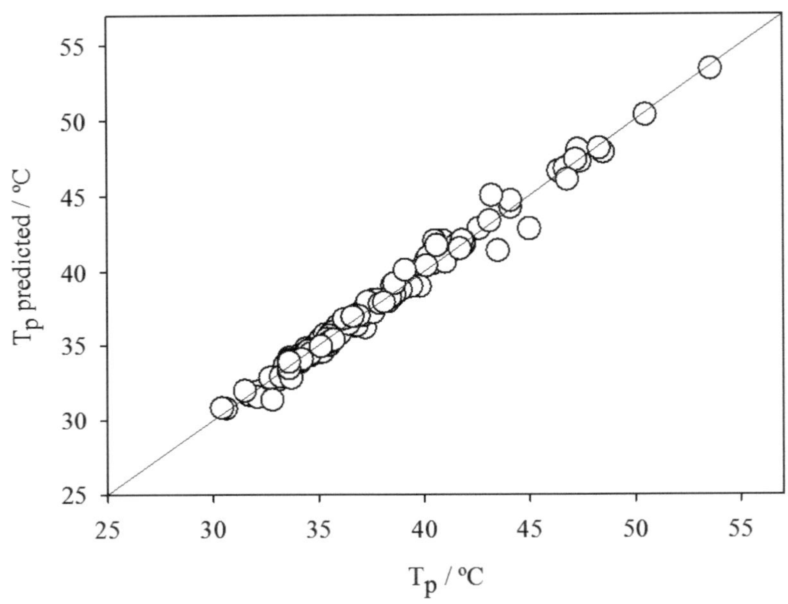

Figure 13. Experimental versus calculated value of T_p for AOT-based microemulsions with carboxylic acids as additive. Modified with permission from Moldes et al. [65].

The input variables used for the n-alkylamine model were as follows: (i) additive concentration, (ii) log *P*, (iii) pK_a, (iv) hydrocarbon chain length and (v) molecular mass of the additive [60]. Different neural models were developed with different 55 amine experimental cases, where 42 experimental cases were used to train the models and 13 experimental cases were used to validate the model [60]. The best neural model presents a topology with five input nodes, two intermediate layers (with 15 and 10 nodes) and one node in the output layer to `predict percolation temperature [60]. This model presents a root-mean-squared error of

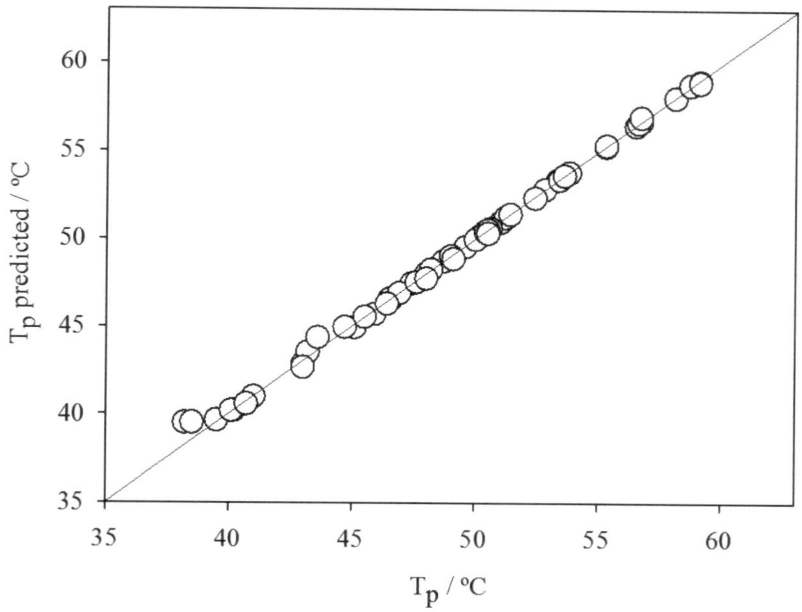

Figure 14. Experimental versus calculated value of T_p for AOT-based microemulsions with alkylamines as additive. Modified with permission from Montoya et al. [60].

0.08°C, with correlation coefficient of 0.9999, for the training phase, and an RMSE of 0.54°C, with R of 0.9976 for the validation phase, see **Figure 14** [60].

4. Conclusions

To summarise, we have demonstrated that ANNs are useful tools for percolation phenomena prediction. Unfortunately, at the moment, we are not able to design a single neural model architecture for additive effect on percolation. There is no doubt that it will be necessary to improve the number of families of molecules used as additives in the design of new models. This way, a single satisfactory model, which is able to predict the behaviour of different additives in a microemulsion system, will be possible.

Acknowledgements

Dr. G. Astray thanks Consellería de Cultura, Educación e Ordenación Universitaria, for the Postdoctoral grant (Plan I2C), P.P.0000 421S 140.08. Dr. A. Cid acknowledges the post-doctoral grant SFRH/BD/78849/2011 and Pest-C/EQB/LA0006/2013 granted to Requimte, both from the Portuguese Foundation for Science and Technology. The authors thank to European Union for FEDER grant.

Author details

Gonzalo Astray[1], Antonio Cid[1,2], Oscar Adrián Moldes[1] and Juan Carlos Mejuto[1*]

*Address all correspondence to: xmejuto@uvigo.es

1 Physical Chemistry Department, Faculty of Science, University of Vigoy, Ourense, Spain

2 UCIBIO, REQUIMTE, Chemistry Department, Faculty of Science and Technology, New University of Lisbon, Caparica, Portugal

References

[1] *Moulik, S.P. and Paul, B.K.*: Adv. Colloid Inteface Sci. 78 (1998) 99.

[2] *Chatterjee, S., Mitra, R.K., Paul, B.K. and Bhattacharya, S.C.*: J. Colloid Interface Sci. 298 (2006) 935.

[3] *Naouli, N., Rosano, H.L. and Kanouni, M.*: J. Dispersion Sci. Technol. 32 (2011) 359.

[4] *Cid, A., Gómez-Díaz, D., Mejuto, J.C. and Navaza, J.M.*: Tenside Surf. Det. 48 (2011) 165.

[5] *García-Río, L., Leis, J.R., Mejuto, J.C., Peña, M.E. and Iglesias, E.*: Langmuir 10 (1994) 1676.

[6] *García-Río, L. and Hervella, P.*: New J. Chem. 31 (2007) 860.

[7] *Jada, A., Lang, J. and Zana, R.*: J. Phys. Chem. 93 (1989) 10.

[8] *Jada, A., Lang, J., Zana, R., Makhloufi, R., Hirsch, E. and Candau, S.J.*: J Phys. Chem. 94 (1990) 387.

[9] *Lang, J., Mascolo, G., Zana, R. and Luisi, P.L.*: J. Phys. Chem. 94 (1990) 3069.

[10] *Mathew, C., Patanjali, P. K., Nabi, A. and Maitra, A.*: Colloids Surf. 30 (1988) 253.

[11] *Mitra, R.K. and Paul, B.K.*: J. Colloid Inteface Sci. 291 (2005) 550.

[12] *Paul, B.K. and Mitra, R.K.*: J. Colloid Interface Sci. 295 (2006) 230

[13] *De, M., Bhattacharya, S.C., Panda, A.K. and Moulik, S.P.*: J. Dispersion Sci. Technol. 30 (2009) 277.

[14] *Fanun, M.*: J. Colloid Interface Sci. 343 (2010) 496.

[15] *García-Río, L., Hervés, P., Leis, J.R. and Mejuto, J.C.*: Langmuir 13 (1997) 6083.

[16] *García-Río, L., Hervés, P., Mejuto, J.C., Pérez-Juste, J. and Rodríguez-Dafonte, P.*: J. Colloid Interface Sci. 225 (2000) 259.

[17] *Dasilva-Carvalhal, J., García-Río, L., Gómez-Díaz, D., Mejuto, J.C., and Rodríguez-Dafonte, P.*: Langmuir 19 (2003) 5975.

[18] *Dasilva-Carvalhal, J., García-Río, L., Gómez-Díaz, D., Mejuto, J.C. and Pérez-Lorenzo, M.*: J. Colloid Interface Sci. 292 (2005) 591.

[19] *García-Río, L., Mejuto, J.C., Pérez-Lorenzo, M., Rodríguez–Álvarez, A. and Rodríguez-Dafonte, P.*: Langmuir 21 (2005) 6259.

[20] *Dasilva-Carvalhal, J., Fernández-Gándara, D., García-Río, L. and Mejuto, J.C.*: J. Colloid Interface Sci. 301 (2006) 637.

[21] *Cid-Samamed, A., García-Río, L., Fernández-Gandara, D., Mejuto, J.C., Morales, J. and Pérez-Lorenzo, M.*: J. Colloid Inteface Sci. 318 (2008) 525.

[22] *Arias-Barros, S.I., Cid, A., García-Río, L., Mejuto, J.C. and Morales, J.*: Colloid Polym. Sci. 288 (2010) 217.

[23] *Desai, K.M., Survase, S.A., Saudagar, P.S., Lele, S.S. and Singhal, R.S.*: Biochem. Eng. J. 41 (2008) 266.

[24] *Astray, G., Gullón, B., Labidi, J. and Gullón,* P.: Ind. Crop. Prod. 92 (2016) 290.

[25] *Shoaib, M., Shamseldin, A.Y., Melville, B.W. and Khan, M.M.*: J. Hydrol. 535 (2016) 211.

[26] *Sánchez-Mesa, J.A., Galan, C., Martínez-Heras, J.A. and Hervás-Martínez, C.*: Clin. Exp. Allergy 32 (2002) 1606.

[27] *Astray, G., Soto, B., Lopez, D., Iglesias, M.A. and Mejuto, J.C.*: Water Sci. Tech. 73 (2016) 1756.

[28] *Zarei, K. and Atabati, M.*: J. Solution Chem. 42 (2013) 516.

[29] *Ren, Y., Suganthan, P.N. and Srikanth, N.*: IEEE Trans. Sustainable Energy 6 (2015) 236.

[30] *Zeinalizadeh, N., Shojaie, A.A. and Shariatmadari, M.*: Mark. Intell. Planning 33 (2015) 717.

[31] *Csépe, Z., Makra, L., Voukantsis, D., Matyasovszky, I., Tusnády, G., Karatzas, K. and Thibaudon, M.*: Sci. Total Environ. 476 (2014) 542.

[32] *Babic, B.R., Nešic, N. and Miljkovic, Z.*: Artif. Intell. Eng. Des. Anal. Manuf. 25 (2011) 289.

[33] *Kim, M.W. and Huang, J.S.*: Phys. Rev. A. 34 (1986) 719.

[34] *Hait, S.K., Moulik, S.P. and Palepu, R.*: Langmuir 18 (2002) 2471.

[35] *Astray, G., Fernández-González, M., Rodríguez-Rajo, F.J., López, D. and Mejuto, J.C.*: Sci Total Environ 548, (2016) 110.

[36] *Govindaraju, R.S.*: J. Hydrol. Eng. 5 (2000) 115.

[37] *Banakar, A. and Azeem, M.F.*: App. Soft Comp. J. 8 (2008) 1463.

[38] *Kriesel, D.* A Brief Introduction to Neural Networks (2007) available at http://www.dkriesel.com

[39] *Moldes, O.A., Mejuto, J.C., Rial-Otero, R. and Simal-Gandara, J.*: Crit. Rev. Food Sci. Nutr. (2015) DOI: 10.1080/10408398.2015.1078277

[40] *Burkert, U., and Allinger, N.L.* "Molecular Mechanics"; ASC: Washington, DC. (1982).

[41] *Allinger, N.L.*: J. Am. Chem. Soc. 99 (1977) 8127.

[42] *Cid, A., Astray, G., Manso, J.A., Mejuto, J.C. and Moldes, O.A.*: Tenside Surf. Det. 48 (2011) 477.

[43] *Alvarez, E., García-Río, L., Mejuto, J.C. and Navaza, J.M.*: J. Chem. Eng. Data 43 (1998) 519.

[44] *Alvarez, E., García-Río, L., Mejuto, J.C., Navaza, J.M. and Pérez-Juste, J.*: J. Chem. Eng. Data 44 (1999) 850.

[45] *Alvarez, E., García-Río, L., Mejuto, J.C., Navaza, J.M. and Pérez-Juste, J.*: J. Chem. Eng. Data 44 (1999) 846.

[46] *Cid, A., Moldes, O.A., Morales, J. and Mejuto, J.C.*: J. Appl. Sol. Chem. Model. 3 (2014) 106.

[47] *Montoya, I.A., Astray, G., Cid, A., Manso, J.A., Moldes, O.A. and Mejuto, J.C.*: Tenside Surf. Det. 49 (2012) 316.

[48] *Álvarez, E., García-Río, L., Leis, J.R., Mejuto, J.C. and Navaza, J.M.*: J. Chem. Eng. Data 43 (1998) 123.

[49] *Moldes, O.A., Cid, A., Astray, G. and Mejuto, J.C.* Tenside Surf. Det. 51 (2014) 533.

[50] *Dasilva-Carbalhal, J., García-Río, L., Gómez-Díaz, D., Mejuto, J.C. and Navaza, J.M.*: J. Chem. Eng. Data, 51 (2006) 1749.

[51] *Álvarez, E., García-Río, L., Gómez-Díaz, D., Mejuto, J.C. and Navaza, J.M.*: J. Chem. Eng. Data, 46 (2001) 526.

[52] *Dasilva-Carbalhal, J., García-Río, L., Gómez-Díaz, D., Mejuto, J.C. and Navaza, J.M.*: J. Chem. Eng. Data, 50 (2005) 529.

[53] *Dasilva-Carbalhal, J., Fernández-Gándara, D., García-Río, L. and Mejuto, J.C.*: J. Colloid Interface Sci. 301 (2006) 637.

[54] *Dasilva-Carbalhal, J., García-Río, L., Gómez-Díaz, D., Mejuto, J.C. and Ridríguez-Dafonte, P.*: Langmuir, 19 (2003) 5975.

[55] *Moldes, O.A., Cid, A., Montoya, I.A. and Mejuto J.C.*: Tenside Surf. Det. 52 (2015) 264.

[56] *Dasilva-Carbalhal, J., García-Río, L., Gómez-Díaz, D., Mejuto, J.C. and Pérez-Lorenzo, M.*: J. Colloid Interface Sci. 292 (2005) 591.

[57] *Álvarez, E., García-Río, L., Mejuto, J.C. and Navaza, J.M.*: J. Chem. Eng. Data 44 (1999) 484.

[58] *van Bommel, A., MacIsaac, G., Livingstone, N. and Palepu, R.*: Fluid Phase Eq. 237 (2005) 59.

[59] *Moldes, O.A., Morales, J., Cid, A., Astray, G., Montoya, I.A. and Mejuto, J.C.*: J. Mol. Liq. 215 (2016) 18.

[60] *Montoya, I.A., Moldes, O.A., Cid, A., Astray, G., Gálvez, J.F. and Mejuto, J.C.*: Tenside Surf. Det. 52 (2015) 473.

[61] *Alvarez, E., García-Río, L., Mejuto, J.C. and Navaza, J.M.*: J. Chem. Eng. Data 43 (1998) 433.

[62] *Alvarez, E., García-Río, L., Mejuto, J.C., Navaza, J.M., Pérez-Juste, J. and Rodríguez-Dafonte, P.*: J. Chem. Eng. Data 44 (1999) 1286.

[63] *García-Río, L., Hervés, P., Mejuto, J.C., Pérez-Juste, J. and Rodríguez-Dafonte, P.*: J. Colloid Interface Sci. 225 (2000) 259.

[64] *Cid, A., García-Río, L., Fernández-Gándara, D., Mejuto J.C., Morales, J. and Pérez-Lorenzo, M.*: J. Colloid Interface Sci. 318 (2008) 525.

[65] *Moldes, O.A., Astray, G., Cid, A., Iglesias-Otero, M.A., Morales, J. and Mejuto J.C.*: Tenside Surf. Det. 50 (2013) 360.